AF413116

Published by:
Kea Books, Ann Arbor, MI.
www.keabooks.org.

```
BibTeX information
@book{milne2022,
  author={Milne, J. S.},
  title={Fields and Galois Theory},
  year={2022},
  publisher={Kea Books},
  address={Ann Arbor, MI}
}
```


A pdf file of this book is available for download at
`https://www.jmilne.org/math/Books`.

Thanks to Bill Casselman for help with the cover.

The Kea is a friendly intelligent parrot found only in the mountains of New Zealand.

Contents

Notation

We use the standard notation: $\mathbb{N} = \{0, 1, 2, \ldots\}$, $\mathbb{Z} = $ ring of integers, $\mathbb{R} = $ field of real numbers, $\mathbb{C} = $ field of complex numbers, $\mathbb{F}_p = \mathbb{Z}/p\mathbb{Z} = $ field with p elements.

Given an equivalence relation, $[*]$ denotes the equivalence class containing $*$. The cardinality of a set S is denoted by $|S|$ (so $|S|$ is the number of elements in S when S is finite). Let I and A be sets. A family of elements of A indexed by I, denoted by $(a_i)_{i \in I}$, is a function $i \mapsto a_i \colon I \to A$. Throughout the notes, p is a prime number: $p = 2, 3, 5, 7, 11, \ldots$ If σ is an element of a group, $\langle \sigma \rangle$ denotes the subgroup generated by σ.

$$\begin{aligned}
&X \subset Y && X \text{ is a subset of } Y \text{ (not necessarily proper).} \\
&X \overset{\text{def}}{=} Y && \text{indicates that the equality in question is a definition.} \\
&X \approx Y && X \text{ and } Y \text{ are isomorphic.} \\
&X \simeq Y && X \text{ and } Y \text{ are canonically isomorphic or} \\
& && \quad\quad\quad\quad \text{there is a given isomorphism.}
\end{aligned}$$

Following Bourbaki, we require compact spaces to be Hausdorff.

Prerequisites

Undergraduate linear algebra and the ring theory. Group theory, for example, the first six chapters of my notes GT.

References

monnnn Question nnnn on mathoverflow.net.

sxnnnn Question nnnn on math.stackexchange.com.

PARI An open source computer algebra system that can run in your browser. It is freely available at `http://pari.math.u-bordeaux.fr/`.

The following notes (available at `https://www.jmilne.org/math/`).

GT Group Theory, v4.00, 2021.

ANT Algebraic Number Theory, v3.08, 2020.

CA A Primer of Commutative Algebra, v4.03, 2020.

Acknowledgements

I thank the following for providing corrections and comments for earlier versions of this work: Mike Albert, Terezakis Alexios, Carlos Alberto

Ajila Loayza, Lior Bary-Soroker, Maren Baumann, Leendert Bleijenga, Jin Ce, Tommaso Centeleghe, Sergio Chouhy, Demetres Christofides, Antoine Chambert-Loir, Dustin Clausen, Keith Conrad, Daniel Duparc, Hardy Falk, Ralf Goertz, Le Minh Ha, Matin Hajian, Jens Hansen, Kenneth Harris, Albrecht Hess, Tim Holzschuh, Philip Horowitz, Ivan Ip, Trevor Jarvis, Henry Kim, Martin Klazar, Jasper Loy Jiabao, Hongjun Lee, Weiyi Liu, Dmitry Lyubshin, Geir Arne Magnussen, John McKay, Sarah Manski, Georges E. Melki, Courtney Mewton, C Nebula, Shuichi Otsuka, Dmitri Panov, Artem Pelenitsyn, Alain Pichereau, David G. Radcliffe, Roberto La Scala, Chad Schoen, René Schoof, Prem L Sharma, Dror Speiser, Sam Spiro, Bhupendra Nath Tiwari, Mathieu Vienney, Martin Ward (and class), Yervand Yeghiazarian, Xiande Yang, Wei Xu, Weijie Zhang, and others.

CHAPTER 1

Basic Definitions and Results

We require rings to have a 1, which entails that we require homomorphisms to preserve it.

Rings

A ***ring*** is a set R with two binary operations $+$ and $\cdot$ such that

(a) $(R,+)$ is a commutative group;

(b) $\cdot$ is associative, and there exists an element 1_R such that, for all $a \in R$,
$$a \cdot 1_R = a = 1_R \cdot a \ ;$$

(c) the distributive law holds: for all $a, b, c \in R$,
$$(a+b) \cdot c = a \cdot c + b \cdot c$$
$$a \cdot (b+c) = a \cdot b + a \cdot c.$$

We usually omit "$\cdot$" and write 1 for 1_R when this causes no confusion. If $1_R = 0$, then the ring $R = \{0\}$.

A ***subring*** of a ring R is a subset S that contains 1_R and is closed under addition, passage to the negative, and multiplication. It inherits the structure of a ring from that on R.

A ***homomorphism of rings*** $\alpha \colon R \to R'$ is a map such that
$$\alpha(a+b) = \alpha(a) + \alpha(b), \quad \alpha(ab) = \alpha(a)\alpha(b), \quad \alpha(1_R) = 1_{R'}$$

for all $a, b \in R$. A ring R is said to be ***commutative*** if multiplication is commutative,
$$ab = ba \text{ for all } a, b \in R.$$

A commutative ring is said to be an ***integral domain*** if $1_R \neq 0$ and the cancellation law holds for multiplication,

$$ab = ac, a \neq 0, \text{ implies } b = c.$$

An ***ideal*** I in a commutative ring R is a subgroup of $(R, +)$ that is closed under multiplication by elements of R,

$$r \in R, a \in I, \text{ implies } ra \in I.$$

The ideal generated by elements $a_1, \ldots, a_n$ is denoted by $(a_1, \ldots, a_n)$. For example, (a) is the principal ideal aR.

We assume that the reader has some familiarity with the elementary theory of rings. For example, with the field of fractions of an integral domain, and with the quotient R/I of a ring R by an ideal I. This last is an integral domain if and only if I is prime, i.e., $I \neq R$ and $ab \in R$ implies in $a \in R$ or $b \in R$. Also, in $\mathbb{Z}$ (more generally, in any Euclidean domain) an ideal I is generated by any "smallest" nonzero element of I, and unique factorization into powers of irreducible elements holds. We write $\gcd(a, b)$ for the greatest common divisor of a and b, e.g., $\gcd(a, 0) = a$. An element of an integral domain is irreducible if it is neither zero nor a unit and admits only trivial factorizations, and it is prime if it is nonzero and generates a prime ideal — in a unique factorization domain, the two notions coincide.

Fields

DEFINITION 1.1 A ***field*** is a set F with binary operations $+$ and $\cdot$ such that
  (a) $(F, +)$ is a commutative group;
  (b) $(F^\times, \cdot)$, where $F^\times \stackrel{\text{def}}{=} F \smallsetminus \{0\}$, is a commutative group;
  (c) the distributive law holds.

Thus, a field is a nonzero commutative ring such that every nonzero element has an inverse. In particular, it is an integral domain. A field contains at least two distinct elements, 0 and 1. The smallest, and one of the most important, fields is $\mathbb{F}_2 = \mathbb{Z}/2\mathbb{Z} = \{0, 1\}$.

A ***subfield*** S of a field F is a subring that is closed under passage to the inverse. It inherits the structure of a field from that on F.

LEMMA 1.2 *A nonzero commutative ring R is a field if and only if it has no ideals other than (0) and R.*

PROOF. Suppose that R is a field, and let I be a nonzero ideal in R. If a is a nonzero element of I, then $1 = a^{-1}a \in I$, and so $I = R$. Conversely, suppose that R is a commutative ring with no proper nonzero ideals. If $a \neq 0$, then $(a) = R$, and so there exists a b in R such that $ab = 1$. $\qquad\square$

EXAMPLE 1.3 The following are fields: $\mathbb{Q}, \mathbb{R}, \mathbb{C}, \mathbb{F}_p$ (p prime).

A ***homomorphism of fields*** is simply a homomorphism of rings. Such a homomorphism is always injective, because its kernel is a proper ideal (it does not contain 1), which must therefore be zero.

Let F be a field. An ***F-algebra*** (or ***algebra over*** F) is a ring R containing F as a subring. A ***homomorphism of F-algebras*** $\alpha \colon R \to R'$ is a homomorphism of rings such that $\alpha(c) = c$ for every $c \in F$. An F-algebra R is ***finite*** if it is finite-dimensional as a F-vector space.

The characteristic of a field

One checks easily that the map

$$\mathbb{Z} \to F, \quad n \mapsto n \cdot 1_F \overset{\text{def}}{=} 1_F + 1_F + \cdots + 1_F \quad (n \text{ copies of } 1_F),$$

is a homomorphism of rings. For example,

$$\underbrace{(1_F + \cdots + 1_F)}_{m} + \underbrace{(1_F + \cdots + 1_F)}_{n} = \underbrace{1_F + \cdots + 1_F}_{m+n}$$

because of the associativity of addition. Therefore its kernel is an ideal in $\mathbb{Z}$.

CASE 1: The kernel of the map is (0), so that

$$n \cdot 1_F = 0 \quad (\text{in } F) \implies n = 0 \quad (\text{in } \mathbb{Z}).$$

Nonzero integers map to invertible elements of F under $n \mapsto n \cdot 1_F \colon \mathbb{Z} \to F$, and so this map extends to a homomorphism

$$\frac{m}{n} \mapsto (m \cdot 1_F)(n \cdot 1_F)^{-1} \colon \mathbb{Q} \hookrightarrow F.$$

In this case, F contains a copy of $\mathbb{Q}$, and we say that it has ***characteristic zero***.

CASE 2: The kernel of the map is $\neq (0)$, so that $n \cdot 1_F = 0$ for some $n \neq 0$. The smallest positive such n is a prime p, because otherwise F would contain two nonzero elements whose product is zero, and p generates the kernel. Thus, the map $n \mapsto n \cdot 1_F : \mathbb{Z} \to F$ defines an isomorphism from $\mathbb{Z}/p\mathbb{Z}$ onto the subring

$$\{m \cdot 1_F \mid m \in \mathbb{Z}\}$$

of F. In this case, F contains a copy of $\mathbb{F}_p$, and we say that it has **characteristic** p.

A field isomorphic to one of the fields $\mathbb{F}_2, \mathbb{F}_3, \mathbb{F}_5, \ldots, \mathbb{Q}$ is called a **prime field**. Every field contains exactly one prime field (as a subfield).

1.4 More generally, a commutative ring R is said to have **characteristic** p (resp. 0) if it contains a prime field of characteristic p (resp. 0) as a subring.[1] Then the prime field is unique and, by definition, contains 1_R. If R has characteristic $p \neq 0$, then $1_R + \cdots + 1_R = 0$ (p terms).

Let R be a nonzero commutative ring. If R has characteristic $p \neq 0$, then

$$pa \stackrel{\text{def}}{=} \underbrace{a + \cdots + a}_{p \text{ terms}} = \underbrace{(1_R + \cdots + 1_R)}_{p \text{ terms}} a = 0a = 0$$

for all $a \in R$. Conversely, if $pa = 0$ for all $a \in R$, then R has characteristic p.

Let R be a nonzero commutative ring. The usual argument by induction shows that the binomial theorem holds in R,

$$(a + b)^m = a^m + \binom{m}{1} a^{m-1} b + \binom{m}{2} a^{m-2} b^2 + \cdots + b^m.$$

If p is prime, then it divides

$$\binom{p}{r} \stackrel{\text{def}}{=} \frac{p!}{r!(p-r)!}$$

for all r with $1 \leq r \leq p - 1$ because it divides the numerator but not the denominator. Therefore, when R has characteristic p,

$$(a + b)^p = a^p + b^p \quad \text{for all } a, b \in R,$$

and so the map $a \mapsto a^p : R \to R$ is a homomorphism of rings (even of $\mathbb{F}_p$-algebras). It is called the **Frobenius endomorphism** of R. The map

---

[1] A commutative ring has a characteristic if and only if it contains a field as a subring. For example, neither $\mathbb{Z}$ nor $\mathbb{F}_2 \times \mathbb{F}_3$ has a characteristic.

$a \mapsto a^{p^n} \colon R \to R$, $n \geq 1$, is the composite of n copies of the Frobenius endomorphism, and so it also is a homomorphism. Therefore,

$$(a_1 + \cdots + a_m)^{p^n} = a_1^{p^n} + \cdots + a_m^{p^n}$$

for all $a_i \in R$.

When F is a field, the Frobenius endomorphism is injective, and hence is an automorphism if F is finite.

The ***characteristic exponent*** of a field F is 1 if F has characteristic 0, and p if F has characteristic $p \neq 0$. Thus, if q is the characteristic exponent of F and $n \geq 1$, then $x \mapsto x^{q^n}$ is an isomorphism of F onto a subfield of F (denoted F^{q^n}).

Review of polynomial rings

Let F be a field.

1.5 The ring $F[X]$ of polynomials in the symbol (or "indeterminate" or "variable") X with coefficients in F is an F-vector space with basis 1, X, $\ldots$, X^n, $\ldots$, and with the multiplication

$$\left(\sum_i a_i X^i \right) \left(\sum_j b_j X^j \right) = \sum_k \left(\sum_{i+j=k} a_i b_j \right) X^k.$$

The F-algebra $F[X]$ has the following universal property: for any F-algebra R and element r of R, there is a unique homomorphism of F-algebras $\alpha \colon F[X] \to R$ such that $\alpha(X) = r$.

1.6 ***Division algorithm***: given $f(X)$, $g(X) \in F[X]$ with $g \neq 0$, there exist $q(X)$, $r(X) \in F[X]$ with $r = 0$ or $\deg(r) < \deg(g)$ such that

$$f = gq + r;$$

moreover, $q(X)$ and $r(X)$ are uniquely determined. Thus $F[X]$ is a Euclidean domain with deg as norm, and so it is a unique factorization domain. A polynomial in $F[X]$ is ***irreducible*** if it is nonconstant and not the product of two polynomials of lower degree.

1.7 Let $f \in F[X]$ be nonconstant, and let $a \in F$. The division algorithm shows that

$$f = (X - a)q + c$$

with $q \in F[X]$ and $c \in F$. Therefore, if a is a root of f (that is, $f(a) = 0$), then $X - a$ divides f. From unique factorization, it now follows that f has at most $\deg(f)$ roots (see also Exercise 1-3).

1.8 **Euclid's algorithm**: Let $f(X), g(X) \in F[X]$. Euclid's algorithm constructs polynomials $a(X), b(X)$, and $d(X)$ such that

$$a(X) \cdot f(X) + b(X) \cdot g(X) = d(X), \quad \deg(a) < \deg(g), \quad \deg(b) < \deg(f),$$

and $d(X) = \gcd(f, g)$.

Recall how it goes. We may assume that $\deg(f) \geq \deg(g)$ since the argument is the same in the opposite case. Using the division algorithm, we construct a sequence of quotients and remainders

$$f = q_0 g + r_0$$
$$g = q_1 r_0 + r_1$$
$$r_0 = q_2 r_1 + r_2$$
$$\cdots$$
$$r_{n-2} = q_n r_{n-1} + r_n$$
$$r_{n-1} = q_{n+1} r_n$$

with r_n the last nonzero remainder. Then, r_n divides r_{n-1}, hence $r_{n-2}, \ldots$, hence g, and hence f. Moreover,

$$r_n = r_{n-2} - q_n r_{n-1} = r_{n-2} - q_n (r_{n-3} - q_{n-1} r_{n-2}) = \cdots = af + bg$$

and so every common divisor of f and g divides r_n: we have shown that $r_n = \gcd(f, g)$.

Let $af + bg = d$. If $\deg(a) \geq \deg(g)$, write $a = gq + r$ with $\deg(r) < \deg(g)$. Then

$$rf + (b + qf)g = d,$$

and $b + qf$ has degree $< \deg(f)$ because $(b + qf)g = d - rf$, which has degree $< \deg(g) + \deg(f)$.

PARI knows how to do Euclidean division: typing `divrem(13,5)` in PARI returns $[2, 3]$, meaning that $13 = 2 \times 5 + 3$, and `gcd(48,87)` returns the greatest common divisor 3 of 48 and 87.

1.9 Let I be a nonzero ideal in $F[X]$, and let f be a nonzero polynomial of least degree in I; then $I = (f)$ (because $F[X]$ is a Euclidean domain). When we choose f to be **monic**, i.e., to have leading coefficient one, it is uniquely determined by I. Thus, there is a one-to-one correspondence between the nonzero ideals of $F[X]$ and the monic polynomials in $F[X]$. The prime ideals correspond to the irreducible monic polynomials.

1.10 As $F[X]$ is an integral domain, we can form its field of fractions $F(X)$. Its elements are quotients f/g, with f and g polynomials, $g \neq 0$, and $f/g = f'/g'$ if and only if $fg' = f'g$.

Factoring polynomials

The following results help in deciding whether a polynomial is reducible, and in finding its factors.

PROPOSITION 1.11 *Let $r \in \mathbb{Q}$ be a root of a polynomial*

$$a_m X^m + a_{m-1} X^{m-1} + \cdots + a_0, \quad a_i \in \mathbb{Z},$$

and write $r = c/d$, $c, d \in \mathbb{Z}$, $\gcd(c, d) = 1$. Then $c \,|\, a_0$ and $d \,|\, a_m$.

PROOF. It is clear from the equation

$$a_m c^m + a_{m-1} c^{m-1} d + \cdots + a_0 d^m = 0$$

that $d \,|\, a_m c^m$, and therefore, $d \,|\, a_m$. Similarly, $c \,|\, a_0$. $\quad\square$

EXAMPLE 1.12 The polynomial $f(X) = X^3 - 3X - 1$ is irreducible in $\mathbb{Q}[X]$ because its only possible roots are ± 1, and $f(1) \neq 0 \neq f(-1)$.

PROPOSITION 1.13 (GAUSS'S LEMMA) *Let $f(X) \in \mathbb{Z}[X]$. If $f(X)$ factors nontrivially in $\mathbb{Q}[X]$, then it factors nontrivially in $\mathbb{Z}[X]$.*

PROOF. Let $f = gh$ in $\mathbb{Q}[X]$ with g, h nonconstant. For suitable integers m and n, $g_1 \overset{\text{def}}{=} mg$ and $h_1 \overset{\text{def}}{=} nh$ have coefficients in $\mathbb{Z}$, and so we have a factorization

$$mnf = g_1 \cdot h_1 \quad \text{in } \mathbb{Z}[X].$$

If a prime number p divides mn, then, looking modulo p, we obtain an equation

$$0 = \overline{g}_1 \cdot \overline{h}_1 \quad \text{in } \mathbb{F}_p[X].$$

Since $\mathbb{F}_p[X]$ is an integral domain, this implies that p divides all the coefficients of one of the polynomials g_1, h_1, say g_1, so that $g_1 = pg_2$ for some $g_2 \in \mathbb{Z}[X]$. Thus, we have a factorization

$$(mn/p)f = g_2 \cdot h_1 \text{ in } \mathbb{Z}[X].$$

Continuing in this fashion, we eventually remove all the prime factors of mn, and so obtain a nontrivial factorization of f in $\mathbb{Z}[X]$. $\square$

PROPOSITION 1.14 *If $f \in \mathbb{Z}[X]$ is monic, then every monic factor of f in $\mathbb{Q}[X]$ lies in $\mathbb{Z}[X]$.*

PROOF. Let g be a monic factor of f in $\mathbb{Q}[X]$, so that $f = gh$ with $h \in \mathbb{Q}[X]$ also monic. Let m, n be positive integers, chosen to have the fewest prime factors, such that $mg, nh \in \mathbb{Z}[X]$. As in the proof of Gauss's Lemma, if a prime p divides mn, then it divides all the coefficients of one of the polynomials mg, nh, say mg, in which case it divides m because g is monic. Now $\frac{m}{p} g \in \mathbb{Z}[X]$, which contradicts the definition of m. $\square$

ASIDE 1.15 We sketch an alternative proof of Proposition 1.14. A complex number α is said to be an ***algebraic integer*** if it is a root of a monic polynomial in $\mathbb{Z}[X]$. Proposition 1.11 shows that every algebraic integer in $\mathbb{Q}$ lies in $\mathbb{Z}$. The algebraic integers form a subring of $\mathbb{C}$ — see Theorem 6.5 of my notes on Commutative Algebra. Now let $\alpha_1, \ldots, \alpha_m$ be the roots of f in $\mathbb{C}$. By definition, they are algebraic integers, and the coefficients of any monic factor of f are polynomials in (certain of) the α_i, and therefore are algebraic integers. If they lie in $\mathbb{Q}$, then they lie in $\mathbb{Z}$.

PROPOSITION 1.16 (EISENSTEIN'S CRITERION) *Let*

$$f = a_m X^m + a_{m-1} X^{m-1} + \cdots + a_0, \quad a_i \in \mathbb{Z};$$

suppose that there is a prime number p such that:

    $\diamond$ *p does not divide a_m,*

    $\diamond$ *p divides $a_{m-1}, \ldots, a_0$,*

    $\diamond$ *p^2 does not divide a_0.*

Then f is irreducible in $\mathbb{Q}[X]$.

PROOF. If $f(X)$ factors nontrivially in $\mathbb{Q}[X]$, then it factors nontrivially in $\mathbb{Z}[X]$, say,

$$a_m X^m + a_{m-1} X^{m-1} + \cdots + a_0 = (b_r X^r + \cdots + b_0)(c_s X^s + \cdots + c_0)$$

with $b_i, c_i \in \mathbb{Z}$ and $r, s < m$. Since p, but not p^2, divides $a_0 = b_0 c_0$, p must divide exactly one of b_0, c_0, say, b_0. Now from the equation

$$a_1 = b_0 c_1 + b_1 c_0,$$

we see that $p | b_1$, and from the equation

$$a_2 = b_0 c_2 + b_1 c_1 + b_2 c_0,$$

that $p|b_2$. By continuing in this way, we find that p divides $b_0, b_1, \ldots, b_r$, which contradicts the condition that p does not divide a_m. $\square$

The last three propositions hold *mutatis mutandis* with $\mathbb{Z}$ replaced by a unique factorization domain R (replace $\mathbb{Q}$ with the field of fractions of R and p with an irreducible element of R).

REMARK 1.17 There is an algorithm for factoring a polynomial in $\mathbb{Q}[X]$. To see this, consider $f \in \mathbb{Q}[X]$. Multiply $f(X)$ by a rational number so that it is monic, and then replace it by $D^{\deg(f)} f(\frac{X}{D})$, with D equal to a common denominator for the coefficients of f, to obtain a monic polynomial with integer coefficients. Thus we need consider only polynomials

$$f(X) = X^m + a_1 X^{m-1} + \cdots + a_m, \quad a_i \in \mathbb{Z}.$$

From the fundamental theorem of algebra (see Theorem 5.6 below), we know that f splits completely in $\mathbb{C}[X]$,

$$f(X) = \prod_{i=1}^{m} (X - \alpha_i), \quad \alpha_i \in \mathbb{C}.$$

From the equation

$$0 = f(\alpha_i) = \alpha_i^m + a_1 \alpha_i^{m-1} + \cdots + a_m,$$

it follows that $|\alpha_i|$ is less than some bound depending only on the degree and coefficients of f; in fact,

$$|\alpha_i| \leq \max\{1, mB\}, \ B = \max |a_i|.$$

Now if $g(X)$ is a monic factor of $f(X)$, then its roots in $\mathbb{C}$ are certain of the α_i, and its coefficients are symmetric polynomials in its roots (see p. 99). Therefore, the absolute values of the coefficients of $g(X)$ are bounded in terms of the degree and coefficients of f. Since they are also integers (by 1.14), we see that there are only finitely many possibilities for $g(X)$. Thus, to find the factors of $f(X)$ we (better PARI) only have to make a finite search.[2]

We shall not concern ourselves with the problem of factoring polynomials in $\mathbb{Q}[X]$ or $\mathbb{F}_p[X]$ because PARI knows how to do this. For example, typing `content(6*X^2+18*X-24)` in PARI returns 6, and typing `factor(6*X^2+18*X-24)` returns $X - 1$ and $X + 4$, showing that

$$6X^2 + 18X - 24 = 6(X - 1)(X + 4) \quad \text{in } \mathbb{Q}[X].$$

---

[2]Of course, there are much faster methods. For example, the Berlekamp–Zassenhaus algorithm factors the polynomial over certain suitable finite fields $\mathbb{F}_p$, lifts the factorizations to rings $\mathbb{Z}/p^m\mathbb{Z}$ for some m, and then searches for factorizations in $\mathbb{Z}[X]$ with the correct form modulo p^m. See the Wikipedia.

Typing `factormod(X^2+3*X+3,7)` returns $X + 4$ and $X + 6$, showing that

$$X^2 + 3X + 3 = (X + 4)(X + 6) \quad \text{in } \mathbb{F}_7[X].$$

REMARK 1.18 One other observation is useful. Let $f \in \mathbb{Z}[X]$. If the leading coefficient of f is not divisible by a prime p, then a nontrivial factorization $f = gh$ in $\mathbb{Z}[X]$ will give a nontrivial factorization $\bar{f} = \bar{g}\bar{h}$ in $\mathbb{F}_p[X]$. Thus, if $f(X)$ is irreducible in $\mathbb{F}_p[X]$ for some prime p not dividing its leading coefficient, then it is irreducible in $\mathbb{Z}[X]$.

This test is very useful, but it is not always effective: for example, $X^4 - 10X^2 + 1$ is irreducible in $\mathbb{Z}[X]$ but it is reducible modulo every prime p. We prove this using only that the product of two nonsquares in $\mathbb{F}_p^\times$ is a square, which follows from the fact that $\mathbb{F}_p^\times$ is cyclic (see Exercise 1-3). If p is such that 2 is a square in $\mathbb{F}_p$, then

$$X^4 - 10X^2 + 1 = (X^2 - 2\sqrt{2}X - 1)(X^2 + 2\sqrt{2}X - 1).$$

If p is such that 3 is a square in $\mathbb{F}_p$, then

$$X^4 - 10X^2 + 1 = (X^2 - 2\sqrt{3}X + 1)(X^2 + 2\sqrt{3}X + 1).$$

If neither 2 nor 3 is a square in $\mathbb{F}_p$, then 6 is a square in $\mathbb{F}_p$, and

$$X^4 - 10X^2 + 1 = (X^2 - (5 + 2\sqrt{6}))(X^2 - (5 - 2\sqrt{6})).$$

The general study of such polynomials requires nonelementary methods. See, for example, the paper Brandl, Amer. Math. Monthly, **93** (1986), pp. 286–288, which proves that for every composite integer $n \geq 1$, there exists a polynomial in $\mathbb{Z}[X]$ of degree n that is irreducible over $\mathbb{Z}$ but reducible modulo all primes.

Extensions

Let F be a field. An ***extension*** of F is field containing F as a subfield. In other words, an extension is an F-algebra whose underlying ring is a field. An extension E of F is, in particular, an F-vector space, whose dimension is called the ***degree*** $[E:F]$ of E over F. An extension is said to be ***finite*** (resp. ***quadratic***, ***cubic***, etc.) if its degree is finite (resp. 2, 3, etc.).

When E and E' are extensions of F, an F-***homomorphism*** $E \to E'$ is a homomorphism $\varphi \colon E \to E'$ such that $\varphi(c) = c$ for all $c \in F$. An F-***isomorphism*** is a bijective F-homomorphism.

EXAMPLE 1.19 (a) The field of complex numbers $\mathbb{C}$ has degree 2 over $\mathbb{R}$ (basis $\{1,i\}$).

(b) The field of real numbers $\mathbb{R}$ has infinite degree over $\mathbb{Q}$: the field $\mathbb{Q}$ is countable, and so every finite-dimensional $\mathbb{Q}$-vector space is also countable, but a famous argument of Cantor shows that $\mathbb{R}$ is not countable.

(c) The field of **Gaussian numbers**

$$\mathbb{Q}(i) \overset{\text{def}}{=} \{a + bi \in \mathbb{C} \mid a, b \in \mathbb{Q}\}$$

has degree 2 over $\mathbb{Q}$ (basis $\{1,i\}$).

(d) The field $F(X)$ has infinite degree over F; in fact, even its subspace $F[X]$ has infinite dimension over F (basis $1, X, X^2, \ldots$).

PROPOSITION 1.20 (MULTIPLICATIVITY OF DEGREES) *Consider fields $L \supset E \supset F$. Then L/F is of finite degree if and only if L/E and E/F are both of finite degree, in which case*

$$[L:F] = [L:E][E:F].$$

PROOF. If L is finite over F, then it is certainly finite over E; moreover, E, being a subspace of a finite-dimensional F-vector space, is also finite-dimensional.

Thus, assume that L/E and E/F are of finite degree, and let $(e_i)_{1 \leq i \leq m}$ be a basis for E as an F-vector space and let $(l_j)_{1 \leq j \leq n}$ be a basis for L as an E-vector space. We'll complete the proof by showing that $(e_i l_j)_{1 \leq i \leq m, 1 \leq j \leq n}$ is a basis for L over F.

First, $(e_i l_j)_{i,j}$ spans L. Let $\gamma \in L$. Then, because $(l_j)_j$ spans L as an E-vector space,

$$\gamma = \sum_j \alpha_j l_j, \qquad \text{some } \alpha_j \in E,$$

and because $(e_i)_i$ spans E as an F-vector space,

$$\alpha_j = \sum_i a_{ij} e_i, \qquad \text{some } a_{ij} \in F.$$

On putting these together, we find that

$$\gamma = \sum_{i,j} a_{ij} e_i l_j.$$

Second, $(e_i l_j)_{i,j}$ is linearly independent. A linear relation $\sum a_{ij} e_i l_j = 0$, $a_{ij} \in F$, can be rewritten $\sum_j (\sum_i a_{ij} e_i) l_j = 0$. The linear independence of the l_j now shows that $\sum_i a_{ij} e_i = 0$ for each j, and the linear independence of the e_i shows that each $a_{ij} = 0$. $\qquad\square$

The subring generated by a subset

An intersection of subrings of a ring is again a ring (this is easy to prove). Let F be a subfield of a field E and S a subset of E. The intersection of all the subrings of E containing F and S is obviously the smallest subring of E containing both F and S. We call it the subring of E **generated by F and S** (or the **F-algebra generated by** S), and we denote it by $F[S]$. When $S = \{\alpha_1, ..., \alpha_n\}$, we write $F[\alpha_1, ..., \alpha_n]$ for $F[S]$. For example, $\mathbb{C} = \mathbb{R}[\sqrt{-1}]$.

LEMMA 1.21 *The ring $F[S]$ consists of the elements of E that can be expressed as finite sums of the form*

$$\sum a_{i_1 \cdots i_n} \alpha_1^{i_1} \cdots \alpha_n^{i_n}, \quad a_{i_1 \cdots i_n} \in F, \quad \alpha_i \in S, \quad i_j \in \mathbb{N}. \tag{1}$$

PROOF. Let R be the set of all such elements. Obviously, R is a subring of E containing F and S and contained in every other such subring. Therefore it equals $F[S]$. $\qquad\square$

EXAMPLE 1.22 The ring $\mathbb{Q}[\pi]$, $\pi = 3.14159...$, consists of the real numbers that can be expressed as a finite sum

$$a_0 + a_1\pi + a_2\pi^2 + \cdots + a_n\pi^n, \quad a_i \in \mathbb{Q}.$$

The ring $\mathbb{Q}[i]$ consists of the complex numbers of the form $a + bi$, $a, b \in \mathbb{Q}$.

Note that the expression of an element in the form (1) will *not* be unique in general. This is so already in $\mathbb{R}[i]$.

LEMMA 1.23 *Let R be a finite F-algebra. If R is an integral domain, then it is a field.*

PROOF. Let α be a nonzero element of R — we have to show that α has an inverse in R. The map $x \mapsto \alpha x : R \to R$ is an injective linear map of finite-dimensional F-vector spaces, and is therefore surjective. In particular, there is an element $\beta \in R$ such that $\alpha\beta = 1$. $\qquad\square$

In particular, every subring (containing F) of a finite extension of F is a field.

The subfield generated by a subset

An intersection of subfields of a field is again a field. Let F be a subfield of a field E and S a subset of E. The intersection of all the subfields of E containing F and S is obviously the smallest subfield of E containing both F and S. We call it the subfield of E *generated by F and S* (or *generated over F by S*), and we denote it by $F(S)$. It is the field of fractions of $F[S]$ in E because this is a subfield of E containing F and S and contained in every other such field. When $S = \{\alpha_1,...,\alpha_n\}$, we write $F(\alpha_1,...,\alpha_n)$ for $F(S)$. Thus, $F[\alpha_1,...,\alpha_n]$ consists of all elements of E that can be expressed as polynomials in the α_i with coefficients in F, and $F(\alpha_1,...,\alpha_n)$ consists of all elements of E that can be expressed as a quotient of two such polynomials.

Lemma 1.23 shows that $F[S]$ is already a field if it is finite-dimensional over F, in which case $F(S) = F[S]$.

EXAMPLE 1.24 (a) The field $\mathbb{Q}(\pi)$, $\pi = 3.14\ldots$, consists of the complex numbers that can be expressed as a quotient

$$g(\pi)/h(\pi), \quad g(X), h(X) \in \mathbb{Q}[X], \quad h(X) \neq 0.$$

(b) The ring $\mathbb{Q}[i]$ is already a field.

An extension E of F is said to be *simple* if $E = F(\alpha)$ some $\alpha \in E$. For example, $\mathbb{Q}(\pi)$ and $\mathbb{Q}[i]$ are simple extensions of $\mathbb{Q}$.

Let F and F' be subfields of a field E. The intersection of the subfields of E containing both F and F' is obviously the smallest subfield of E containing both F and F'. We call it the *composite* of F and F' in E, and we denote it by $F \cdot F'$. It can also be described as the subfield of E generated over F by F', or the subfield generated over F' by F:

$$F(F') = F \cdot F' = F'(F).$$

Construction of some extensions

Let $f(X) \in F[X]$ be a monic polynomial of degree m, and let (f) be the ideal generated by $f(X)$. Consider the quotient ring $F[X]/(f)$, and write x for the image of X in $F[X]/(f)$, i.e., x is the coset $X + (f(X))$.

(a) The map

$$P(X) \mapsto P(x) \colon F[X] \to F[x]$$

is an F-homomorphism sending $f(X)$ to 0. Therefore, $f(x) = 0$.

(b) The division algorithm shows that every element g of $F[X]/(f)$ is represented by a unique polynomial r of degree $< m$. Hence each element of $F[x]$ can be expressed uniquely as a sum

$$a_0 + a_1 x + \cdots + a_{m-1} x^{m-1}, \qquad a_i \in F. \tag{2}$$

(c) To add two elements, expressed in the form (2), simply add the corresponding coefficients.

(d) To multiply two elements expressed in the form (2), multiply in the usual way, and use the relation $f(x) = 0$ to express the monomials of degree $\geq m$ in x in terms of lower degree monomials.

(e) *Now assume that $f(X)$ is irreducible.* Then every nonzero $\alpha \in F[x]$ has an inverse, which can be found as follows. Use (b) to write $\alpha = g(x)$ with $g(X)$ a polynomial of degree $\leq m - 1$, and apply Euclid's algorithm in $F[X]$ to find polynomials $a(X)$ and $b(X)$ such that

$$a(X)f(X) + b(X)g(X) = d(X)$$

with $d(X)$ the gcd of f and g. In our case, $d(X)$ is 1 because $f(X)$ is irreducible and $\deg g(X) < \deg f(X)$. When we replace X with x, the equality becomes

$$b(x)g(x) = 1.$$

Hence $b(x)$ is the inverse of $g(x)$.

We have proved the following statement.

1.25 *For a monic irreducible polynomial $f(X)$ of degree m in $F[X]$,*

$$F[x] \overset{\text{def}}{=} F[X]/(f(X))$$

is a field of degree m over F. Computations in $F[x]$ come down to computations in F.

Note that, because $F[x]$ is a field, $F(x) = F[x]$.[3]

EXAMPLE 1.26 Let $f(X) = X^2 + 1 \in \mathbb{R}[X]$. Then $\mathbb{R}[x]$ has
    elements: $a + bx, a, b \in \mathbb{R}$;
    addition: $(a + bx) + (a' + b'x) = (a + a') + (b + b')x$;
    multiplication: $(a + bx)(a' + b'x) = (aa' - bb') + (ab' + a'b)x$;
    inverses: in this case, it is possible write down the inverse of $a + bx$
directly.
We usually write i for x and $\mathbb{C}$ for $\mathbb{R}[x]$.

---

[3]Thus, we can denote it by $F(x)$ or by $F[x]$. The former is more common, but I use $F[x]$ to emphasize the fact that its elements are polynomials in x.

EXAMPLE 1.27 Let $f(X) = X^3 - 3X - 1 \in \mathbb{Q}[X]$. We observed in (1.12) that this is irreducible over $\mathbb{Q}$, and so $\mathbb{Q}[x]$ is a field. It has basis $\{1, x, x^2\}$ as a $\mathbb{Q}$-vector space. Let

$$\beta = x^4 + 2x^3 + 3 \in \mathbb{Q}[x].$$

Using that $x^3 - 3x - 1 = 0$, we find that $\beta = 3x^2 + 7x + 5$. Because $X^3 - 3X - 1$ is irreducible,

$$\gcd(X^3 - 3X - 1, 3X^2 + 7X + 5) = 1.$$

In fact, Euclid's algorithm gives

$$\left(X^3 - 3X - 1\right)\left(\tfrac{-7}{37}X + \tfrac{29}{111}\right) + \left(3X^2 + 7X + 5\right)\left(\tfrac{7}{111}X^2 - \tfrac{26}{111}X + \tfrac{28}{111}\right) = 1.$$

Hence

$$\left(3x^2 + 7x + 5\right)\left(\tfrac{7}{111}x^2 - \tfrac{26}{111}x + \tfrac{28}{111}\right) = 1,$$

and we have found the inverse of β.

We can also do this in PARI: b=Mod(X^4+2*X^3+3,X^3-3*X-1) reveals that $\beta = 3x^2 + 7x + 5$ in $\mathbb{Q}[x]$, and b^(-1) reveals that $\beta^{-1} = \tfrac{7}{111}x^2 - \tfrac{26}{111}x + \tfrac{28}{111}$.

Stem fields

Let f be a monic irreducible polynomial in $F[X]$. A pair (E, α) consisting of an extension E of F and an $\alpha \in E$ is called[4] a **stem field for** f if $E = F[\alpha]$ and $f(\alpha) = 0$. For example, the pair (E, α) with $E = F[X]/(f) = F[x]$ and $\alpha = x$ is a stem field for f. Let (E, α) be a stem field, and consider the surjective homomorphism of F-algebras

$$g(X) \mapsto g(\alpha) \colon F[X] \to E.$$

Its kernel is generated by a nonzero monic polynomial, which divides f, and so must equal f. Therefore the homomorphism defines an F-isomorphism

$$x \mapsto \alpha \colon F[x] \to E, \quad \text{where } F[x] = F[X]/(f).$$

---

[4]Following A.A. Albert (*Modern Higher Algebra*, 1937) who calls the splitting field of a polynomial its root field.

In other words, the stem field (E,α) of f is F-isomorphic to the standard stem field $(F[X]/(f),x)$. It follows that every element of a stem field (E,α) for f can be written uniquely in the form

$$a_0 + a_1\alpha + \cdots + a_{m-1}\alpha^{m-1}, \quad a_i \in F, \quad m = \deg(f),$$

and that arithmetic in $F[\alpha]$ can be performed using the same rules as in $F[x]$. If (E',α') is a second stem field for f, then there is a unique F-isomorphism $E \to E'$ sending α to α'. We sometimes abbreviate "stem field $(F[\alpha],\alpha)$" to "stem field $F[\alpha]$".

Algebraic and transcendental elements

Let F be a field and E an integral domain containing F as a subring. An element α of E defines a homomorphism

$$f(X) \mapsto f(\alpha) \colon F[X] \to E.$$

There are two possibilities.

CASE 1: The kernel of the map is (0), so that, for $f \in F[X]$,

$$f(\alpha) = 0 \implies f = 0 \text{ (in } F[X]).$$

In this case, we say that α **transcendental over** F. The homomorphism $X \mapsto \alpha \colon F[X] \to F[\alpha]$ is an isomorphism, and it extends to an isomorphism $F(X) \to F(\alpha)$ if E is a field.

CASE 2: The kernel is $\neq (0)$, so that $g(\alpha) = 0$ for some nonzero $g \in F[X]$. In this case, we say that α is **algebraic over** F. The polynomials g such that $g(\alpha) = 0$ form a nonzero ideal in $F[X]$, which is generated by the monic polynomial f of least degree such $f(\alpha) = 0$. We call f the **minimal** (or **minimum**) **polynomial** of α over F.[5] It is irreducible, because otherwise there would be two nonzero elements of E whose product is zero.

The minimal polynomial is characterized as an element of $F[X]$ by each of the following conditions,

    $\diamond$ f is monic, $f(\alpha) = 0$, and f divides every other g in $F[X]$ such that $g(\alpha) = 0$;

    $\diamond$ f is the monic polynomial of least degree such that $f(\alpha) = 0$;

---

[5]When we order the polynomials by degree, f is a minimal element of the set of polynomials having α as a root, and the minimum (i.e., least) element of the the set of *monic* polynomials having α as a root.

⋄ f is monic, irreducible, and $f(\alpha) = 0$.

Note that $g(X) \mapsto g(\alpha)$ defines an isomorphism $F[X]/(f) \to F[\alpha]$. Since the first is a field, so also is the second. Thus, $F[\alpha]$ is a stem field for f.

EXAMPLE 1.28 Let $\alpha \in \mathbb{C}$ be such that $\alpha^3 - 3\alpha - 1 = 0$. Then $X^3 - 3X - 1$ is monic, irreducible, and has α as a root, and so it is the minimal polynomial of α over $\mathbb{Q}$. The set $\{1, \alpha, \alpha^2\}$ is a basis for $\mathbb{Q}[\alpha]$ over $\mathbb{Q}$. The calculations in Example 1.27 show that if β is the element $\alpha^4 + 2\alpha^3 + 3$ of $\mathbb{Q}[\alpha]$, then $\beta = 3\alpha^2 + 7\alpha + 5$, and

$$\beta^{-1} = \tfrac{7}{111}\alpha^2 - \tfrac{26}{111}\alpha + \tfrac{28}{111}.$$

REMARK 1.29 PARI knows how to compute in $\mathbb{Q}[a]$. For example, typing `factor(X^4+4)` returns the factorization

$$X^4 + 4 = (X^2 - 2X + 2)(X^2 + 2X + 2)$$

in $\mathbb{Q}[X]$. Now type `F=nfinit(a^2+2*a+2)` to define a number field "F" generated over $\mathbb{Q}$ by a root a of $X^2 + 2X + 2$. Then `nffactor(F,x^4+4)` returns the factorization

$$X^4 + 4 = (X - a - 2)(X - a)(X + a))(X + a + 2),$$

in $\mathbb{Q}[a]$.

An extension E of F is said to be **algebraic** (and E is said to be **algebraic over** F), if every element of E is algebraic over F; otherwise it is said to be **transcendental** (and E is said to be **transcendental over** F). Thus, E/F is transcendental if at least one element of E is transcendental over F.

PROPOSITION 1.30 *Let $E \supset F$ be fields. If E/F is finite, then E is algebraic and finitely generated (as a field) over F; conversely, if E is generated over F by a finite set of algebraic elements, then it is of finite degree over F.*

PROOF. $\implies$: To say that an element α of E is transcendental over F amounts to saying that its powers $1, \alpha, \alpha^2, \ldots$ are linearly independent over F. As E is finite over F, its elements are algebraic over F. It remains to show that E is finitely generated over F. If $E = F$, then it is generated by the empty set. Otherwise, there exists an $\alpha_1 \in E \smallsetminus F$. If $E \neq F[\alpha_1]$, then there exists an $\alpha_2 \in E \smallsetminus F[\alpha_1]$, and so on. Since

$$[F[\alpha_1] : F] < [F[\alpha_1, \alpha_2] : F] < \cdots < [E : F]$$

this process terminates with $E = F[\alpha_1, \alpha_2, \ldots, \alpha_n]$.

$\Longleftarrow$: Let $E = F(\alpha_1, \ldots, \alpha_n)$ with $\alpha_1, \alpha_2, \ldots \alpha_n$ algebraic over F. The extension $F(\alpha_1)/F$ is finite because α_1 is algebraic over F, and the extension $F(\alpha_1, \alpha_2)/F(\alpha_1)$ is finite because α_2 is algebraic over F and hence over $F(\alpha_1)$. Thus, by (1.20), $F(\alpha_1, \alpha_2)$ is finite over F. Now repeat the argument. $\qquad\square$

COROLLARY 1.31 *Consider fields $L \supset E \supset F$. If L is algebraic over E and E is algebraic over F, then L is algebraic over F.*

PROOF. By assumption, every $\alpha \in L$ is a root of a monic polynomial

$$X^m + a_{m-1} X^{m-1} + \cdots + a_0, \quad a_i \in E.$$

According to the proposition, the ring $F[a_0, \ldots, a_{m-1}]$ is finite over F and $F[a_0, \ldots, a_{m-1}, \alpha]$ is finite over $F[a_0, \ldots, a_{m-1}]$, and so $F[a_0, \ldots, a_{m-1}, \alpha]$ is finite over F (see 1.20); hence α is algebraic over F. $\qquad\square$

PROPOSITION 1.32 *Let F be a field and R an integral domain containing F as a subring. If R is generated as an F-algebra by elements algebraic over F, then it is a field algebraic over F.*

PROOF. Suppose first that $R = F[\alpha_1, \ldots, \alpha_n]$ with each α_i algebraic over F. For each i, there exist an $m_i > 0$ and $a_j \in F$ such that

$$\alpha_i^{m_i} = a_0 + \cdots + a_{m_i-1} \alpha_i^{m_i-1}.$$

Hence R is spanned as an F-vector space by the elements

$$\alpha_1^{i_1} \cdots \alpha_n^{i_n}, \quad i_1 < m_1, \ldots, i_n < m_n.$$

In particular, R is a finite F-algebra, and hence a field algebraic over F (1.23, 1.30). In the general case, each element of R is contained in the F-algebra generated by a finite set of elements algebraic over F, and so it has an inverse in R and is algebraic over F. $\qquad\square$

Transcendental numbers

A complex number is said to be **algebraic** or **transcendental** according as it is algebraic or transcendental over $\mathbb{Q}$. First we provide a little history.

1844: Liouville showed that certain numbers, now called Liouville numbers, are transcendental.

1873: Hermite showed that e is transcendental.

1874: Cantor showed that the set of algebraic numbers is countable, but that $\mathbb{R}$ is not countable, and so most real numbers are transcendental.

1882: Lindemann showed that π is transcendental.

1934: Gel'fond and Schneider independently showed that α^{β} is transcendental if α and β are algebraic, $\alpha \neq 0, 1$, and $\beta \notin \mathbb{Q}$. (This was the seventh of Hilbert's famous problems.)

2022: Euler's constant

$$\gamma \overset{\text{def}}{=} \lim_{n \to \infty} \left(1 + \frac{1}{2} + \cdots + \frac{1}{n} - \log n \right)$$

has not yet been proven to be transcendental or even irrational (see Lagarias, *Euler's constant.* Bull. Amer. Math. Soc. 50 (2013), 527–628, arXiv:1303:1856).

2022: The numbers $e + \pi$ and $e - \pi$ are surely transcendental, but they have not even been proved to be irrational!

PROPOSITION 1.33 *The set of algebraic numbers is countable.*

PROOF. Every algebraic number is a root of a polynomial

$$a_0 X^n + a_1 X^{n-1} + \cdots + a_n, \quad a_0, \ldots, a_n \in \mathbb{Z}.$$

For an $N \in \mathbb{N}$, there are only finitely many such polynomials with $n \leq N$ and $|a_0|, \ldots, |a_n| \leq N$, and each polynomial has only finitely many roots. Thus, the set of algebraic numbers is a countable union of finite sets $\bigcup_{N \geq 1} A(N)$, and any such union is countable — for example, choose a bijection from some segment $[0, n(1)]$ of $\mathbb{N}$ onto $A(1)$, extend it to a bijection from a segment $[0, n(2)]$ onto $A(2)$, and so on. $\square$

A typical Liouville number is $\sum_{n=0}^{\infty} \frac{1}{10^{n!}}$ — in its decimal expansion there are increasingly long strings of zeros. Since its decimal expansion is not periodic, the number is not rational. We prove that the analogue of this number in base 2 is transcendental.

THEOREM 1.34 (LIOUVILLE) *The number* $\alpha = \sum \frac{1}{2^{n!}}$ *is transcendental.*

PROOF. [6]Suppose not, and let

$$f(X) = X^d + a_1 X^{d-1} + \cdots + a_d, \quad a_i \in \mathbb{Q},$$

---

[6]This proof, which I learnt from David Masser, also works for $\sum \frac{1}{a^{n!}}$, where a is any integer ≥ 2.

be the minimal polynomial of α over $\mathbb{Q}$. Thus $[\mathbb{Q}[\alpha]:\mathbb{Q}] = d$. Choose a nonzero integer D such that $D \cdot f(X) \in \mathbb{Z}[X]$.

Let $\Sigma_N = \sum_{n=0}^{N} \frac{1}{2^{n!}}$, so that $\Sigma_N \to \alpha$ as $N \to \infty$, and let $x_N = f(\Sigma_N)$. As α is not rational, $f(X)$, being irreducible of degree > 1, has no rational root. Since $\Sigma_N \neq \alpha$, it cannot be a root of $f(X)$, and so $x_N \neq 0$. Obviously, $x_N \in \mathbb{Q}$; in fact $(2^{N!})^d D x_N \in \mathbb{Z}$, and so

$$|(2^{N!})^d D x_N| \geq 1. \tag{3}$$

From the fundamental theorem of algebra (see 5.6 below), we know that f splits in $\mathbb{C}[X]$, say,

$$f(X) = \prod_{i=1}^{d} (X - \alpha_i), \quad \alpha_i \in \mathbb{C}, \quad \alpha_1 = \alpha,$$

and so

$$|x_N| = \prod_{i=1}^{d} |\Sigma_N - \alpha_i| \leq |\Sigma_N - \alpha_1|(\Sigma_N + M)^{d-1},$$

where $M = \max_{i \neq 1}\{1, |\alpha_i|\}$. But

$$|\Sigma_N - \alpha_1| = \sum_{n=N+1}^{\infty} \frac{1}{2^{n!}} \leq \frac{1}{2^{(N+1)!}} \left(\sum_{n=0}^{\infty} \frac{1}{2^n} \right) = \frac{2}{2^{(N+1)!}}.$$

Hence

$$|x_N| \leq \frac{2}{2^{(N+1)!}} \cdot (\Sigma_N + M)^{d-1}$$

and

$$|(2^{N!})^d D x_N| \leq 2 \cdot \frac{2^{d \cdot N!} D}{2^{(N+1)!}} \cdot (\Sigma_N + M)^{d-1}$$

which tends to 0 as $N \to \infty$ because $\frac{2^{d \cdot N!}}{2^{(N+1)!}} = \left(\frac{2^d}{2^{N+1}} \right)^{N!} \to 0$. This contradicts (3). $\qquad\square$

Constructions with straight-edge and compass.

The Greeks understood integers and the rational numbers. They were surprised to find that the length of the diagonal of a square of side 1, namely, $\sqrt{2}$, is not rational. They thus realized that they needed to extend their number system. They then hoped that the "constructible" numbers would suffice.

Suppose that we are given a length, which we call 1, a straight-edge, and a compass (device for drawing circles). A real number (better a length) is **constructible** if it can be constructed by forming successive intersections of

⋄ lines drawn through two points already constructed, and

⋄ circles with centre a point already constructed and radius a constructed length.

This led them to three famous questions that they were unable to answer: is it possible to duplicate the cube, trisect an angle, or square the circle by straight-edge and compass constructions? We'll see that the answer to all three is negative.

Let F be a subfield of $\mathbb{R}$. For a positive $a \in F$, $\sqrt{a}$ denotes the positive square root of a in $\mathbb{R}$. The F-**plane** is $F \times F \subset \mathbb{R} \times \mathbb{R}$. We make the following definitions:

An F-**line** is a line in $\mathbb{R} \times \mathbb{R}$ through two points in the F-plane. These are the lines given by equations

$$ax + by + c = 0, \quad a,b,c \in F.$$

An F-**circle** is a circle in $\mathbb{R} \times \mathbb{R}$ with centre an F-point and radius an element of F. These are the circles given by equations

$$(x-a)^2 + (y-b)^2 = c^2, \quad a,b,c \in F.$$

LEMMA 1.35 *Let $L \neq L'$ be F-lines, and let $C \neq C'$ be F-circles.*

(a) *$L \cap L' = \emptyset$ or consists of a single F-point.*

(b) *$L \cap C = \emptyset$ or consists of one or two points in the $F[\sqrt{e}]$-plane, some $e \in F, e > 0$.*

(c) *$C \cap C' = \emptyset$ or consists of one or two points in the $F[\sqrt{e}]$-plane, some $e \in F, e > 0$.*

PROOF. The points in the intersection are found by solving the simultaneous equations, and hence by solving (at worst) a quadratic equation with coefficients in F.   □

LEMMA 1.36 *(a) If c and d are constructible, then so also are $c + d$, $-c$, cd, and $\frac{c}{d}$ $(d \neq 0)$.*

(b) If $c > 0$ is constructible, then so also is $\sqrt{c}$.

SKETCH OF PROOF. First show that it is possible to construct a line perpendicular to a given line through a given point, and then a line parallel to a given line through a given point. Hence it is possible to construct a triangle similar to a given one on a side with given length. By an astute choice of the triangles, one constructs cd and c^{-1}. For (b), draw a circle of radius $\frac{c+1}{2}$ and centre $(\frac{c+1}{2}, 0)$, and draw a vertical line through the point $A = (1,0)$ to meet the circle at P. The length AP is $\sqrt{c}$. (For more details, see Michael Artin, *Algebra*, 1991, Chap. 13, Section 4.)   □

THEOREM 1.37   (a) *The set of constructible numbers is a field.*

  (b) *A number α is constructible if and only if it is contained in a subfield of $\mathbb{R}$ of the form*

$$\mathbb{Q}[\sqrt{a_1}, \ldots, \sqrt{a_r}], \quad a_i \in \mathbb{Q}[\sqrt{a_1}, \ldots, \sqrt{a_{i-1}}], \quad a_i > 0.$$

PROOF. (a) This restates (a) of Lemma 1.36.

  (b) It follows from Lemma 1.35 that every constructible number is contained in such a field $\mathbb{Q}[\sqrt{a_1}, \ldots, \sqrt{a_r}]$. Conversely, if, for some $i < r$, the elements of $\mathbb{Q}[\sqrt{a_1}, \ldots, \sqrt{a_{i-1}}]$ are constructible, then $\sqrt{a_i}$ is constructible (by 1.36b), and so the elements of $\mathbb{Q}[\sqrt{a_1}, \ldots, \sqrt{a_i}]$ are constructible (by (a)). Applying this for $i = 0, 1, \ldots$, we find that the elements of $\mathbb{Q}[\sqrt{a_1}, \ldots, \sqrt{a_r}]$ are constructible.   □

COROLLARY 1.38 *If α is constructible, then α is algebraic over $\mathbb{Q}$, and $[\mathbb{Q}[\alpha]:\mathbb{Q}]$ is a power of 2.*

PROOF. According to Proposition 1.20, $[\mathbb{Q}[\alpha]:\mathbb{Q}]$ divides

$$[\mathbb{Q}[\sqrt{a_1}] \cdots [\sqrt{a_r}]:\mathbb{Q}]$$

and $[\mathbb{Q}[\sqrt{a_1}, \ldots, \sqrt{a_r}]:\mathbb{Q}]$ is a power of 2.   □

COROLLARY 1.39 *It is impossible to duplicate the cube by straight-edge and compass constructions.*

PROOF. The problem is to construct a cube with volume 2. This requires that we construct the real root of the polynomial $X^3 - 2$. But this polynomial is irreducible (by Eisenstein's criterion 1.16), and so $[\mathbb{Q}[\sqrt[3]{2}]:\mathbb{Q}] = 3$.   □

COROLLARY 1.40 *In general, it is impossible to trisect an angle by straight-edge and compass constructions.*

PROOF. Knowing an angle is equivalent to knowing the cosine of the angle. Therefore, to trisect 3θ, we have to construct a solution to

$$\cos 3\theta = 4\cos^3\theta - 3\cos\theta.$$

For example, take $3\theta = 60$ degrees. As $\cos 60° = \frac{1}{2}$, to construct $\cos\theta$, we have to find a root of $8x^3 - 6x - 1$, which is irreducible (apply 1.11), and so $[\mathbb{Q}[\cos\theta]:\mathbb{Q}] = 3$. $\qquad\square$

COROLLARY 1.41 *It is impossible to square the circle by straight-edge and compass constructions.*

PROOF. A square with the same area as a circle of radius r has side $\sqrt{\pi}r$. Since π is transcendental[7], so also is $\sqrt{\pi}$. $\qquad\square$

We next consider another problem that goes back to the ancient Greeks: list the integers n such that the regular n-sided polygon can be constructed using only straight-edge and compass. Here we consider the question for a prime p (see 5.12 for the general case). Note that $X^p - 1$ is not irreducible; in fact

$$X^p - 1 = (X - 1)(X^{p-1} + X^{p-2} + \cdots + 1).$$

LEMMA 1.42 *If p is prime, then $X^{p-1} + \cdots + 1$ is irreducible; hence $\mathbb{Q}[e^{\frac{2\pi i}{p}}]$ has degree $p - 1$ over $\mathbb{Q}$.*

PROOF. Let $f(X) = (X^p - 1)/(X - 1) = X^{p-1} + \cdots + 1$; then

$$f(X + 1) = \frac{(X + 1)^p - 1}{X} = X^{p-1} + \cdots + a_i X^i + \cdots + p,$$

with $a_i = \binom{p}{i+1}$. We know (1.4) that $p|a_i$ for $i = 1, ..., p - 2$, and so the polynomial $f(X + 1)$ is irreducible by Eisenstein's criterion 1.16. This implies that $f(X)$ is irreducible. $\qquad\square$

In order to construct a regular p-gon, p an odd prime, we need to construct

$$\cos\frac{2\pi}{p} = \frac{e^{\frac{2\pi i}{p}} + e^{-\frac{2\pi i}{p}}}{2}.$$

Note that

$$\mathbb{Q}[e^{\frac{2\pi i}{p}}] \supset \mathbb{Q}[\cos\frac{2\pi}{p}] \supset \mathbb{Q}.$$

---

[7]Proofs of this can be found in many books on number theory, for example, in 11.14 of Hardy and Wright, *An Introduction to the Theory of Numbers*, Fourth Edition, Oxford, 1960.

The degree of $\mathbb{Q}[e^{\frac{2\pi i}{p}}]$ over $\mathbb{Q}[\cos\frac{2\pi}{p}]$ is 2 because the equation

$$\alpha^2 - 2\cos\tfrac{2\pi}{p}\cdot\alpha + 1 = 0, \quad \alpha = e^{\frac{2\pi i}{p}},$$

shows that it is at most 2, and it is not 1 because $e^{\frac{2\pi i}{p}} \notin \mathbb{R}$. Hence

$$[\mathbb{Q}[\cos\tfrac{2\pi}{p}]:\mathbb{Q}] = \frac{p-1}{2}.$$

We deduce that, if the regular p-gon is constructible, then $(p-1)/2$ is a power of 2. Later (5.12) we shall prove a converse. Thus, the regular p-gon (p prime) is constructible if and only if $p = 2^n + 1$ for some positive integer n.

A number $2^n + 1$ can be prime only if n is a power of 2, because, otherwise, $n = rs$ with s odd, and

$$Y^s + 1 = (Y + 1)(Y^{s-1} - Y^{s-2} + \cdots + 1)$$
$$2^{rs} + 1 = (2^r + 1)((2^r)^{s-1} - (2^r)^{s-2} + \cdots + 1).$$

We conclude that the primes p for which the regular p-gon is constructible are exactly those of the form $2^{2^r} + 1$ for some r. Such p are called **Fermat primes** (because Fermat conjectured that all numbers of the form $2^{2^r} + 1$ are prime). For $r = 0, 1, 2, 3, 4$, we have $2^{2^r} + 1 = 3, 5, 17, 257, 65537$, which are indeed prime, but Euler showed that $2^{32} + 1 = (641)(6700417)$, and we do not know of any more Fermat primes. It is expected that there are no more, but this has not been proved. See Wikipedia: Fermat number.

Gauss showed that

$$\cos\tfrac{2\pi}{17} = -\tfrac{1}{16} + \tfrac{1}{16}\sqrt{17} + \tfrac{1}{16}\sqrt{34 - 2\sqrt{17}} + \tfrac{1}{8}\sqrt{17 + 3\sqrt{17} - \sqrt{34 - 2\sqrt{17}} - 2\sqrt{34 + 2\sqrt{17}}}$$

when he was 18 years old. This success encouraged him to become a mathematician.

Algebraically closed fields

Let F be a field. A polynomial is said to **split** in $F[X]$ if it is a product of polynomials of degree at most 1 in $F[X]$.

PROPOSITION 1.43 *For a field Ω, the following statements are equivalent:*

   (a) *Every nonconstant polynomial in $\Omega[X]$ splits in $\Omega[X]$.*

(b) *Every nonconstant polynomial in $\Omega[X]$ has at least one root in Ω.*

(c) *The irreducible polynomials in $\Omega[X]$ are those of degree 1.*

(d) *Every field of finite degree over Ω equals Ω.*

PROOF. The implications (a)$\Rightarrow$(b)$\Rightarrow$(c) are obvious.

(c)$\Rightarrow$(a). This follows from the fact that $\Omega[X]$ is a unique factorization domain.

(c)$\Rightarrow$(d). Let E be a finite extension of Ω, and let $\alpha \in E$. The minimal polynomial of α, being irreducible, has degree 1, and so $\alpha \in \Omega$.

(d)$\Rightarrow$(c). Let f be an irreducible polynomial in $\Omega[X]$. Then $\Omega[X]/(f)$ is an extension of Ω of degree $\deg(f)$ (see 1.30), and so $\deg(f) = 1$. □

DEFINITION 1.44 (a) A field Ω is **algebraically closed** if it satisfies the equivalent statements of Proposition 1.43.

(b) A field Ω is an **algebraic closure** of a subfield F if it is algebraically closed and algebraic over F.

For example, the fundamental theorem of algebra (see 5.6 below) says that $\mathbb{C}$ is algebraically closed. It is an algebraic closure of $\mathbb{R}$.

PROPOSITION 1.45 *If Ω is algebraic over F and every polynomial $f \in F[X]$ splits in $\Omega[X]$, then Ω is algebraically closed (and hence an algebraic closure of F).*

PROOF. Let $f = a_n X^n + \cdots + a_0$, $a_i \in \Omega$, be a nonconstant polynomial in $\Omega[X]$. We have to show that f has a root in Ω. We know (see 1.25) that f has a root α in some finite extension Ω' of Ω. Consider the fields

$$F \subset F[a_0, \ldots, a_n] \subset F[a_0, \ldots, a_n, \alpha].$$

Each extension is generated by a finite set of algebraic elements, and hence is finite (1.30). Therefore α lies in a finite extension of F (see 1.20), and so is algebraic over F — it is a root of a polynomial g with coefficients in F. By assumption, g splits in $\Omega[X]$, and so the roots of g in Ω' all lie in Ω. In particular, $\alpha \in \Omega$. □

In fact, it suffices to assume that every $f \in F[X]$ has a root in Ω (see 6.5 below).

PROPOSITION 1.46 *Let F be a field and Ω an integral domain containing F as a subring. Then*

$$\bar{F} \stackrel{\text{def}}{=} \{\alpha \in \Omega \mid \alpha \text{ algebraic over } F\}$$

*is a field (called the **algebraic closure of F in Ω**).*

PROOF. If α and β are algebraic over F, then $F[\alpha,\beta] = F[\alpha][\beta]$ is a field of finite degree over F (see p. 17). Thus, every element of $F[\alpha,\beta]$ is algebraic over F. In particular, $\alpha \pm \beta$, α/β, and $\alpha\beta$ are algebraic over F. $\quad\square$

COROLLARY 1.47 *Let Ω be an algebraically closed field. For any subfield F of Ω, the algebraic closure E of F in Ω is an algebraic closure of F.*

PROOF. It is algebraic over F by definition. Every polynomial in $F[X]$ splits in $\Omega[X]$ and has its roots in E, and so splits in $E[X]$. Now apply Proposition 1.45. $\quad\square$

Thus, when we admit the fundamental theorem of algebra (5.6), every subfield of $\mathbb{C}$ has an algebraic closure (in fact, a canonical algebraic closure). Later (Chapter 6) we'll prove, using the axiom of choice, that every field has an algebraic closure.

NOTES Although various classes of field, for example, number fields and function fields, had been studied earlier, the first systematic account of the theory of abstract fields was given by Steinitz in 1910 (Algebraische Theorie der Körper, J. Reine Angew. Math., 137:167–309). Here he introduced the notion of a prime field, distinguished between separable and inseparable extensions, and showed that every field can be obtained as an algebraic extension of a purely transcendental extension. He also proved that every field has an algebraic closure, unique up to isomorphism. His work influenced later algebraists (Emmy Noether, B. L. van der Waerden, Emil Artin, ...) and his article has been described by Bourbaki as "... a fundamental work that may be considered as having given birth to the current conception[8] of algebra". See: Roquette, Peter, *In memoriam Ernst Steinitz (1871–1928)*. J. Reine Angew. Math. 648 (2010), 1–11.

Exercises

1-1 Let $E = \mathbb{Q}[\alpha]$, where $\alpha^3 - \alpha^2 + \alpha + 2 = 0$. Express the elements $(\alpha^2 + \alpha + 1)(\alpha^2 - \alpha)$ and $(\alpha - 1)^{-1}$ of E in the form $a\alpha^2 + b\alpha + c$ with $a, b, c \in \mathbb{Q}$.

1-2 Determine $[\mathbb{Q}(\sqrt{2}, \sqrt{3}) : \mathbb{Q}]$.

1-3 Let F be a field, and let $f(X) \in F[X]$.

  (a) For every $a \in F$, show that there is a polynomial $q(X) \in F[X]$ such that

$$f(X) = q(X)(X - a) + f(a).$$

---

[8]In which objects are to be defined abstractly by axioms.

(b) Deduce that $f(a) = 0$ if and only if $(X - a) \mid f(X)$.

(c) Deduce that $f(X)$ can have at most $\deg f$ roots.

(d) Let G be a finite abelian group. If G has at most m elements of order dividing m for each divisor m of $(G:1)$, show that G is cyclic.

(e) Deduce that every finite subgroup of $F^\times$, F a field, is cyclic.

1-4 Show that with straight-edge, compass, and angle-trisector, it is possible to construct a regular 7-gon.

1-5 Let $f(X)$ be an irreducible polynomial over F of degree n, and let E be a field extension of F with $[E : F] = m$. If $\gcd(m,n) = 1$, show that f is irreducible over E.

1-6 Show that there does not exist a polynomial $f(X) \in \mathbb{Z}[X]$ of degree > 1 that is irreducible modulo p for all primes p.

1-7 Let $\alpha = \sqrt[3]{2}$, and let R be the set of complex numbers of the form $a + b\alpha + c\alpha^2$ with $a,b,c \in \mathbb{Q}$. Show that R is a field.

1-8 If you understand the Legendre symbol, use its properties to show that the polynomial $(X^2 - 13)(X^2 - 17)(X^2 - 221)$ has roots modulo every integer (but not in $\mathbb{Z}$).

CHAPTER 2

Splitting Fields; Multiple Roots

Homomorphisms from simple extensions.

Let F be a field, and let E and E' be fields containing F. An F-homomorphism $\varphi: E \to E'$ maps a polynomial

$$\sum a_{i_1 \cdots i_m} \alpha_1^{i_1} \cdots \alpha_m^{i_m}, \quad a_{i_1 \cdots i_m} \in F, \quad \alpha_i \in E,$$

to

$$\sum a_{i_1 \cdots i_m} \varphi(\alpha_1)^{i_1} \cdots \varphi(\alpha_m)^{i_m}.$$

An F-homomorphism $E \to E'$ of fields is, in particular, an injective F-linear map of F-vector spaces, and so it is an F-isomorphism if E and E' have the same finite degree over F.

PROPOSITION 2.1 *Let $F(\alpha)$ be a simple extension of F and Ω a second extension of F.*

(a) Suppose that α is transcendental over F. For every F-homomorphism $\varphi: F(\alpha) \to \Omega$, $\varphi(\alpha)$ is transcendental over F, and the map $\varphi \mapsto \varphi(\alpha)$ defines a one-to-one correspondence

$$\{F\text{-homomorphisms } F(\alpha) \to \Omega\} \leftrightarrow \{\text{elements of } \Omega \text{ transcendental over } F\}.$$

(b) Suppose that α is algebraic over F, and let $f(X)$ be its minimal polynomial. For every F-homomorphism $\varphi: F[\alpha] \to \Omega$, $\varphi(\alpha)$ is a root of $f(X)$ in Ω, and the map $\varphi \mapsto \varphi(\alpha)$ defines a one-to-one correspondence

$$\{F\text{-homomorphisms } \varphi: F[\alpha] \to \Omega\} \leftrightarrow \{\text{roots of } f \text{ in } \Omega\}.$$

In particular, the number of such maps is the number of distinct roots of f in the field Ω.

PROOF. (a) To say that α is transcendental over F means that $F[\alpha]$ is isomorphic to the polynomial ring in the symbol α. Therefore, for every $\gamma \in \Omega$, there is a unique F-homomorphism $\varphi \colon F[\alpha] \to \Omega$ such that $\varphi(\alpha) = \gamma$ (see 1.5). This φ extends (uniquely) to the field of fractions $F(\alpha)$ of $F[\alpha]$ if and only if nonzero elements of $F[\alpha]$ are sent to nonzero elements of Ω, which is the case if and only if γ is transcendental over F. Thus we see that there are one-to-one correspondences between (a) the F-homomorphisms $F(\alpha) \to \Omega$, (b) the F-homomorphisms $\varphi \colon F[\alpha] \to \Omega$ such that $\varphi(\alpha)$ is transcendental, (c) the transcendental elements of Ω.

(b) Let $f(X) = \sum a_i X^i$, and consider an F-homomorphism $\varphi \colon F[\alpha] \to \Omega$. On applying φ to the equality $\sum a_i \alpha^i = 0$, we obtain the equality $\sum a_i \varphi(\alpha)^i = 0$, which shows that $\varphi(\alpha)$ is a root of $f(X)$ in Ω. Conversely, if $\gamma \in \Omega$ is a root of $f(X)$, then the map $F[X] \to \Omega$, $g(X) \mapsto g(\gamma)$, factors through $F[X]/(f(X))$. When composed with the inverse of the canonical isomorphism $F[X]/(f(X)) \to F[\alpha]$, this becomes a homomorphism $F[\alpha] \to \Omega$ sending α to γ. $\qquad\square$

EXAMPLE 2.2 Consider a simple algebraic extension $\mathbb{Q}[\alpha]$ of $\mathbb{Q}$, and let f be the minimal polynomial of α. We shall see later that f has $\deg f$ distinct roots in $\mathbb{C}$. Therefore, there are exactly $[\mathbb{Q}[\alpha] \colon \mathbb{Q}]$ distinct $\mathbb{Q}$-homomorphism $\mathbb{Q}[\alpha] \to \mathbb{C}$, each sending α to a root of f in $\mathbb{C}$.

EXAMPLE 2.3 Let F be a field of characteristic $p \neq 0$, and let $a \in F \smallsetminus F^p$. Then $X^p - a$ is irreducible, and we let $F[\alpha]$ be a corresponding stem field. If β is a root of $X^p - a$ in an extension Ω of F, then $X^p - a = (X - \beta)^p$ in $\Omega[X]$, and so there is exactly one F-homomorphism $F[\alpha] \to \Omega$ (sending α to β).

We shall need a slight generalization of 2.1.

PROPOSITION 2.4 *Let $F(\alpha)$ be a simple extension of F and $\varphi_0 \colon F \to \Omega$ a homomorphism from F into a second field Ω.*

(a) If α is transcendental over F, then the map $\varphi \mapsto \varphi(\alpha)$ defines a one-to-one correspondence between extensions $\varphi \colon F(\alpha) \to \Omega$ of φ_0 and elements of Ω transcendental over $\varphi_0(F)$.

(b) If α is algebraic over F, with minimal polynomial $f(X)$, then the map $\varphi \mapsto \varphi(\alpha)$ defines a one-to-one correspondence between extensions $\varphi \colon F(\alpha) \to \Omega$ of φ_0 and roots of $\varphi_0 f$ in Ω. In particular, the number of such maps is the number of distinct roots of $\varphi_0 f$ in Ω.

By $\varphi_0 f$ we mean the polynomial obtained by applying φ_0 to the coefficients of f. By an extension of φ_0 to $F(\alpha)$ we mean a homomorphism $\varphi\colon F(\alpha) \to \Omega$ whose restriction to F is φ_0. The proof of the proposition is essentially the same as that of the preceding proposition (indeed, it is essentially the same proposition).

Splitting fields

Let f be a polynomial with coefficients in F. A field E containing F is said to **split** f if f splits in $E[X]$, i.e.,

$$f(X) = a \prod_{i=1}^{m} (X - \alpha_i) \quad a \in F, \quad \alpha_i \in E.$$

If E splits f and is generated by the roots of f, then it is called a **splitting** or **root field** for f.

Note that $\prod f_i(X)^{m_i}$ $(m_i \geq 1)$ and $\prod f_i(X)$ have the same splitting fields. Note also that f splits in E if it has $\deg(f) - 1$ roots in E because the sum of the roots of f lies in F (if $f = aX^m + a_1 X^{m-1} + \cdots$, then $\sum \alpha_i = -a_1/a$).

EXAMPLE 2.5 Let $f(X) = aX^2 + bX + c \in \mathbb{Q}[X]$, and let $\alpha = \sqrt{b^2 - 4ac}$. The subfield $\mathbb{Q}[\alpha]$ of $\mathbb{C}$ is a splitting field for f.

EXAMPLE 2.6 Let $f(X) = X^3 + aX^2 + bX + c \in \mathbb{Q}[X]$ be irreducible, and let $\alpha_1, \alpha_2, \alpha_3$ be its roots in $\mathbb{C}$. Then $\mathbb{Q}[\alpha_1, \alpha_2, \alpha_3] = \mathbb{Q}[\alpha_1, \alpha_2]$ is a splitting field for $f(X)$. Note that $[\mathbb{Q}[\alpha_1]\colon\mathbb{Q}] = 3$ and that $[\mathbb{Q}[\alpha_1, \alpha_2]\colon\mathbb{Q}[\alpha_1]] = 1$ or 2, and so $[\mathbb{Q}[\alpha_1, \alpha_2]\colon\mathbb{Q}] = 3$ or 6. We'll see later (4.2) that the degree is 3 if and only if the discriminant of $f(X)$ is a square in $\mathbb{Q}$. For example, the discriminant of $X^3 + bX + c$ is $-4b^3 - 27c^2$, and so the splitting field of $X^3 + 10X + 1$ (discriminant -4027) has degree 6 over $\mathbb{Q}$.

PROPOSITION 2.7 *Every polynomial $f \in F[X]$ has a splitting field E_f, and*

$$[E_f\colon F] \leq (\deg f)! \quad \text{(factorial } \deg f).$$

PROOF. Let $F_1 = F[\alpha_1]$ be a stem field for some monic irreducible factor of f in $F[X]$. Then $f(\alpha_1) = 0$, and we let $F_2 = F_1[\alpha_2]$ be a stem field for some monic irreducible factor of $f(X)/(X - \alpha_1)$ in $F_1[X]$. Continuing in this fashion, we arrive at a splitting field E_f. Let $n = \deg f$. Then $[F_1\colon F] = \deg g_1 \leq n$, $[F_2\colon F_1] \leq n - 1, \ldots$, and so $[E_f\colon F] \leq n!$. $\qquad\square$

EXAMPLE 2.8 Let $f(X) = (X^p - 1)/(X - 1) \in \mathbb{Q}[X]$, p prime. If ζ is one root of f, then the remaining roots are $\zeta^2, \zeta^3, \dots, \zeta^{p-1}$, and so the splitting field of f is $\mathbb{Q}[\zeta]$.

EXAMPLE 2.9 Let F have characteristic $p \neq 0$, and let $f = X^p - X - a \in F[X]$. If α is one root of f in some extension of F, then the remaining roots are $\alpha + 1, \dots, \alpha + p - 1$, and so the splitting field of f is $F[\alpha]$.

EXAMPLE 2.10 If α is one root of $X^n - a$, then the remaining roots are all of the form $\zeta\alpha$, where $\zeta^n = 1$. Therefore, $F[\alpha]$ is a splitting field for $X^n - a$ if and only if F contains all the nth roots of 1 (by which we mean that $X^n - 1$ splits in $F[X]$). Note that if p is the characteristic of F, then $X^p - 1 = (X - 1)^p$, and so F automatically contains all the pth roots of 1.

ASIDE 2.11 Let F be a field. For a given integer n, there may or may not exist polynomials of degree n in $F[X]$ whose splitting field has degree $n!$ — this depends on F. For example, there do not exist such polynomials for $n > 1$ if $F = \mathbb{C}$ (see 5.6), nor for $n > 2$ if $F = \mathbb{R}$ or $F = \mathbb{F}_p$ (see 4.22). However, later (4.32) we'll see how to write down infinitely many polynomials of degree n in $\mathbb{Q}[X]$ with splitting fields of degree $n!$.

Homomorphisms of algebraic extensions

PROPOSITION 2.12 *Let $f \in F[X]$. Let E be an extension of F generated by the roots of f in E and Ω an extension of F splitting f. There exists an F-homomorphism $\varphi \colon E \to \Omega$; the number of such homomorphisms is at most $[E:F]$, and equals $[E:F]$ if f has distinct roots in Ω.*

PROOF. We may suppose that f is monic.

By hypothesis, $f = \prod(X - \beta_i)$ in $\Omega[X]$. Let L be a subfield of Ω containing F, and let g be a monic factor of f in $L[X]$. Then g divides f in $\Omega[X]$, and so it is a product there of some of the $X - \beta_i$. In particular, we see that g splits in Ω, and that it has distinct roots in Ω if f does..

By hypothesis, $E = F[\alpha_1, \dots, \alpha_m]$ with each α_i a root of $f(X)$ in E. The minimal polynomial of α_1 is an irreducible polynomial f_1 dividing f. From the initial observation with $L = F$, we see that f_1 splits in Ω, and that its roots are distinct if the roots of f are distinct. According to Proposition 2.1, there exists an F-homomorphism $\varphi_1 \colon F[\alpha_1] \to \Omega$, and the number of such homomorphisms is at most $[F[\alpha_1]:F]$, with equality holding when f has distinct roots in Ω.

The minimal polynomial of α_2 over $F[\alpha_1]$ is an irreducible factor f_2 of f in $F[\alpha_1][X]$. On applying the initial observation with $L = \varphi_1 F[\alpha_1]$ and $g = \varphi_1 f_2$, we see that $\varphi_1 f_2$ splits in Ω, and that its roots are distinct if the roots of f are distinct. According to Proposition 2.4, each φ_1 extends to a homomorphism $\varphi_2 \colon F[\alpha_1, \alpha_2] \to \Omega$, and the number of extensions is at most $[F[\alpha_1, \alpha_2] \colon F[\alpha_1]]$, with equality holding when f has distinct roots in Ω.

On combining these statements we conclude that there exists an F-homomorphism $\varphi \colon F[\alpha_1, \alpha_2] \to \Omega$, and that the number of such homomorphisms is at most $[F[\alpha_1, \alpha_2] \colon F]$, with equality holding if f has distinct roots in Ω.

On repeating this argument m times, we obtain the proposition. $\qquad\square$

COROLLARY 2.13 *If E_1 and E_2 are both splitting fields for f, then every F-homomorphism $E_1 \to E_2$ is an isomorphism. In particular, any two splitting fields for f are F-isomorphic.*

PROOF. Every F-homomorphism $E_1 \to E_2$ is injective, and so, if there exists such a homomorphism, then $[E_1 \colon F] \le [E_2 \colon F]$. If E_1 and E_2 are both splitting fields for f, then 2.12 shows that there exist homomorphisms $E_1 \leftrightarrows E_2$, and so $[E_1 \colon F] = [E_2 \colon F]$. It follows that every F-homomorphism $E_1 \to E_2$ is an F-isomorphism. $\qquad\square$

COROLLARY 2.14 *Let E and L be extension of F, with E finite over F. The number of F-homomorphisms $E \to L$ is at most $[E \colon F]$.*

PROOF. Write $E = F[\alpha_1, \ldots, \alpha_m]$, and let $f \in F[X]$ be the product of the minimal polynomials of the α_i; thus E is generated over F by roots of f. Let Ω be a splitting field for f regarded as an element of $L[X]$. Proposition 2.12 shows that there exists an F-homomorphism $E \to \Omega$, and the number of such homomorphisms is at most $[E \colon F]$. As an F-homomorphism $E \to L$ can be regarded as an F-homomorphism $E \to \Omega$, this proves the corollary.$\square$

REMARK 2.15 Let $E_1, E_2, \ldots, E_m$ be finite extensions of F, and let L be an extension of F. From the corollary we see that there exists a finite extension L_1/L such that L_1 contains an isomorphic image of E_1; then that there exists a finite extension L_2/L_1 such that L_2 contains an isomorphic image of E_2. On continuing in this fashion, we find that there exists a finite extension Ω/L such that Ω contains an isomorphic copy of every E_i.

WARNING 2.16 Let $f \in F[X]$. If E and E' are both splitting fields of f, then we know that there exists an F-isomorphism $E \to E'$, but there will in general be no *preferred* such isomorphism. Error and confusion can result if the fields are simply identified. Also, it makes no sense to speak of "the field $F[\alpha]$ generated by a root of f" unless f is irreducible (the fields generated by the roots of two different factors are unrelated). Even when f is irreducible, it makes no sense to speak of "the field $F[\alpha,\beta]$ generated by two roots α,β of f" (the extensions of $F[\alpha]$ generated by the roots of two different factors of f in $F[\alpha][X]$ may be very different).

Multiplicity of roots

Even when polynomials in $F[X]$ have no common factor in $F[X]$, one might expect that they could acquire a common factor in $\Omega[X]$ for some $\Omega \supset F$. In fact, this does not happen — greatest common divisors do not change when the field is extended.

PROPOSITION 2.17 *Let f and g be polynomials in $F[X]$, and let Ω be an extension of F. If $r(X)$ is the gcd of f and g computed in $F[X]$, then it is also the gcd of f and g in $\Omega[X]$. In particular, distinct monic irreducible polynomials in $F[X]$ do not acquire a common root in any extension of F.*

PROOF. Let $r_F(X)$ and $r_\Omega(X)$ be the greatest common divisors of f and g in $F[X]$ and $\Omega[X]$ respectively. Certainly $r_F(X)|r_\Omega(X)$ in $\Omega[X]$, but Euclid's algorithm (1.8) shows that there are polynomials a and b in $F[X]$ such that

$$a(X)f(X) + b(X)g(X) = r_F(X),$$

and so $r_\Omega(X)$ divides $r_F(X)$ in $\Omega[X]$.

For the second statement, note that the hypotheses imply that $\gcd(f,g) = 1$ (in $F[X]$), and so f and g can not acquire a common factor in any extension field. $\qquad\square$

The proposition allows us to speak of the greatest common divisor of f and g without reference to a field.

Let $f \in F[X]$. Then f splits into linear factors

$$f(X) = a \prod_{i=1}^{r}(X-\alpha_i)^{m_i}, \ a \in F, \ \alpha_i \text{ distinct}, \ m_i \geq 1, \ \sum_{i=1}^{r} m_i = \deg(f),$$

$$(4)$$

in $E[X]$ for some extension E of F (see 2.7). We say that α_i is a root of f of *multiplicity* m_i in E. If $m_i > 1$, then α_i is said to be a *multiple root* of f, and otherwise it is a *simple root*.

The unordered sequence of integers $m_1,\ldots,m_r$ in (4) is independent of the extension E chosen to split f. Certainly, it is unchanged when E is replaced with its subfield $F[\alpha_1,\ldots,\alpha_r]$, and so we may suppose that E is a splitting field for f. Let E and E' be splitting fields for F, and suppose that $f(X) = a \prod_{i=1}^{r}(X - \alpha_i)^{m_i}$ in $E[X]$ and $f(X) = a \prod_{i=1}^{r'}(X - \alpha_i')^{m_i'}$ in $E'[X]$. Let $\varphi: E \to E'$ be an F-isomorphism, which exists by 2.13, and extend it to an isomorphism $E[X] \to E'[X]$ by sending X to X. Then φ maps the factorization of f in $E[X]$ onto a factorization $f(X) = a \prod_{i=1}^{r}(X - \varphi(\alpha_i))^{m_i}$ in $E'[X]$. By unique factorization, this coincides with the earlier factorization in $E'[X]$ up to a renumbering of the α_i. Therefore $r = r'$, and

$$\{m_1,\ldots,m_r\} = \{m_1',\ldots,m_r'\}$$

(equality of multisets).

We say that f **has a multiple root** when at least one of the $m_i > 1$, and that f has **only simple roots** when all $m_i = 1$. Thus "f has a multiple root" means "f has a multiple root in one, hence every, extension of F splitting f".

Separable polynomials

When does a polynomial have a multiple root? If f has a multiple factor in $F[X]$, say $f = g^2 h$, then obviously it will have a multiple root. If is a product of distinct irreducible polynomials, then Proposition 2.17 shows that f has a multiple root if and only if at least one of its factors has a multiple root. Thus, it suffices to determine when an *irreducible* polynomial has a multiple root.

EXAMPLE 2.18 Let F be of characteristic $p \neq 0$, and assume that F contains an element a that is not a pth-power, for example, $a = T$ in the field $\mathbb{F}_p(T)$. Then $X^p - a$ is irreducible in $F[X]$, but $X^p - a = (X - \alpha)^p$ in its splitting field (see 1.4). Thus an irreducible polynomial can have multiple roots.

The derivative of a polynomial $f(X) = \sum a_i X^i$ is defined to be $f'(X) = \sum i a_i X^{i-1}$. The usual rules for differentiating sums and products still hold, but note that in characteristic p the derivative of X^p is zero.

LEMMA 2.19 *A root of f is multiple if and only if it is also a root of f'.*

PROOF. Let

$$f(X) = (X - \alpha)^m g(X), \quad m \geq 1, \quad g(\alpha) \neq 0,$$

in some extension field. Then

$$f'(X) = \begin{cases} g(X) + (X - \alpha)g'(X) & \text{if } m = 1, \\ m(X - \alpha)^{m-1} g(X) + (X - \alpha)^m g'(X) & \text{if } m > 1. \end{cases} \quad (5)$$

Thus $f'(\alpha) = 0 \iff m > 1$. $\qquad\square$

PROPOSITION 2.20 *For a nonconstant irreducible polynomial f in $F[X]$, the following statements are equivalent:*

(a) *f has a multiple root;*

(b) *$\gcd(f, f') \neq 1$;*

(c) *F has nonzero characteristic p and f is a polynomial in X^p;*

(d) *all the roots of f are multiple.*

PROOF. (a) $\Rightarrow$ (b). If α is a multiple root of f, then f and f' have $X - \alpha$ as a common factor.

(b) $\Rightarrow$ (c). As f is irreducible and $\deg(f') < \deg(f)$,

$$\gcd(f, f') \neq 1 \implies f' = 0.$$

Let $f = a_0 + \cdots + a_d X^d$, $d \geq 1$. Then

$$f' = a_1 + \cdots + i a_i X^{i-1} + \cdots + d a_d X^{d-1},$$

which is the zero polynomial if only if F has characteristic $p \neq 0$ and $a_i = 0$ for all i not divisible by p.

(c) $\Rightarrow$ (d). By hypothesis, $f(X) = g(X^p)$ with $g(X) \in F[X]$. Let $g(X) = \prod_i (X - a_i)^{m_i}$ in some extension field. Then each a_i becomes a pth power, say, $a_i = \alpha_i^p$, in some possibly larger extension field. Now

$$f(X) = g(X^p) = \prod_i (X^p - a_i)^{m_i} = \prod_i (X - \alpha_i)^{p m_i}$$

which shows that every root of $f(X)$ has multiplicity at least p.

(d) $\Rightarrow$ (a). Obvious. $\qquad\square$

PROPOSITION 2.21 *The following conditions on a nonzero polynomial $f \in F[X]$ are equivalent:*

(a) *$\gcd(f, f') = 1$ in $F[X]$;*

(b) f *has no multiple roots.*

PROOF. Let Ω be an extension of F splitting f. We know that a root α of f in Ω is multiple if and only if it is also a root of f'.

If $\gcd(f, f') = 1$, then f and f' have no common factor in $\Omega[X]$ (see 2.17). In particular, they have no common root, and so f has no multiple roots.

If f has no multiple roots, then $\gcd(f, f')$ must be the constant polynomial, because otherwise it would have a root in Ω which would then be a common root of f and f'. $\qquad\square$

DEFINITION 2.22 A polynomial is *separable* if it is nonzero and it has no multiple roots.[1]

Constant polynomials are separable, and a nonconstant irreducible polynomial f is separable unless F has characteristic $p \neq 0$ and f is a polynomial in X^p (see 2.20); in particular, f is separable if p does not divide the degree of f. Let $f = \prod f_i$ with f and the f_i monic and the f_i irreducible; then f is separable if and only if the f_i are distinct and separable. If f is separable as a polynomial in $F[X]$, then it is separable as a polynomial in $E[X]$ for every extension E of F.

Perfect fields

DEFINITION 2.23 A field F is *perfect* if it has characteristic zero or it has characteristic p and every every element of F is a pth power.

Equivalently, a field F of characteristic exponent q is perfect if $F = F^q$.

PROPOSITION 2.24 *A field F is perfect if and only if every irreducible polynomial in $F[X]$ is separable.*

PROOF. If F has characteristic zero, the statement is obvious, and so we may suppose F has characteristic $p \neq 0$. If F contains an element a that is not a pth power, then $X^p - a$ is irreducible in $F[X]$ but not separable (see 2.18). Conversely, if every element of F is a pth power, then every polynomial in X^p with coefficients in F is a pth power in $F[X]$,

$$\sum a_i X^{ip} = \left(\sum b_i X^i\right)^p \quad \text{if} \quad a_i = b_i^p,$$

---

[1]This is Bourbaki's definition. Often, for example, in the books of Jacobson and in earlier versions of these notes, a polynomial f is said to be separable if each of its irreducible factors has only simple roots.

and so it is not irreducible.   □

EXAMPLE 2.25   (a) A finite field F is perfect, because the Frobenius endomorphism $a \mapsto a^p \colon F \to F$ is injective and therefore surjective (by counting).

  (b) A field that can be written as a union of perfect fields is perfect. Therefore, every field algebraic over $\mathbb{F}_p$ is perfect.

  (c) Every algebraically closed field is perfect.

  (d) If F_0 has characteristic $p \neq 0$, then $F = F_0(X)$ is not perfect, because X is not a pth power.

ASIDE 2.26 When F is perfect, we shall see (5.1) that every finite extension E/F is simple, i.e., $E = F[\alpha]$ with α a root of a (separable) polynomial $f \in F[X]$ of degree $[E \colon F]$. Thus it follows directly from (2.1b) that, for any extension Ω of F, the number of F-homomorphisms $E \to \Omega$ is $\leq [E \colon F]$, with equality if and only if f splits in Ω. We cannot use this argument here because it would make the exposition circular.

Exercises

2-1 Let F be a field of characteristic $\neq 2$.

  (a) Let E be quadratic extension of F; show that

$$S(E) = \{a \in F^\times \mid a \text{ is a square in } E\}$$

   is a subgroup of $F^\times$ containing $F^{\times 2}$.

  (b) Let E and E' be quadratic extensions of F; show that there exists an F-isomorphism $\varphi \colon E \to E'$ if and only if $S(E) = S(E')$.

  (c) Show that there is an infinite sequence of fields $E_1, E_2, \ldots$ with E_i a quadratic extension of $\mathbb{Q}$ such that E_i is not isomorphic to E_j for $i \neq j$.

  (d) Let p be an odd prime. Show that, up to isomorphism, there is exactly one field with p^2 elements.

2-2 (a) Let F be a field of characteristic p. Show that if $X^p - X - a$ is reducible in $F[X]$, then it splits into distinct factors in $F[X]$.

  (b) For every prime p, show that $X^p - X - 1$ is irreducible in $\mathbb{Q}[X]$.

2-3 Construct a splitting field for $X^5 - 2$ over $\mathbb{Q}$. What is its degree over $\mathbb{Q}$?

2-4 Find splitting fields for the polynomials $X^{p^m} - 1$ and $X^{p^m} - X$ in $\mathbb{F}_p[X]$. What are their degrees over $\mathbb{F}_p$?

2-5 Let $f \in F[X]$, where F is a field of characteristic 0. Let $d(X) = \gcd(f, f')$. Show that $g(X) \stackrel{\text{def}}{=} f(X)d(X)^{-1}$ has the same roots as $f(X)$, and these are all simple roots of $g(X)$.

2-6 Let $f(X)$ be an irreducible polynomial in $F[X]$, where F has characteristic p. Show that $f(X)$ can be written $f(X) = g(X^{p^e})$ where $g(X)$ is irreducible and separable. Deduce that every root of $f(X)$ has the same multiplicity p^e in any splitting field.

CHAPTER 3

The Fundamental Theorem of Galois Theory

In this chapter, we prove the fundamental theorem of Galois theory, which classifies the subfields of the splitting field of a separable polynomial f in terms of the Galois group of f.

Groups of automorphisms of fields

Consider fields $E \supset F$. An F-isomorphism $E \to E$ is called an F-***automorphism*** of E. The F-automorphisms of E form a group, which we denote $\mathrm{Aut}(E/F)$.

EXAMPLE 3.1 (a) There are two obvious automorphisms of $\mathbb{C}$, namely, the identity map and complex conjugation. We'll see later (9.18) that by using the Axiom of Choice we can construct uncountably many more.

(b) Let $E = \mathbb{C}(X)$. A $\mathbb{C}$-automorphism of E sends X to another generator of E over $\mathbb{C}$. It follows from Lemma 9.24 below that these are exactly the elements $\frac{aX+b}{cX+d}$, $ad - bc \neq 0$. Therefore $\mathrm{Aut}(E/\mathbb{C})$ consists of the maps $f(X) \mapsto f\left(\frac{aX+b}{cX+d}\right)$, $ad - bc \neq 0$, and so

$$\mathrm{Aut}(E/\mathbb{C}) \simeq \mathrm{PGL}_2(\mathbb{C}),$$

the group of invertible 2×2 matrices with complex coefficients modulo its centre. Analysts will note that this is the same as the automorphism group of the Riemann sphere. Here is the explanation. The field E of meromorphic functions on the Riemann sphere $\mathbb{P}_\mathbb{C}^1$ consists of the rational functions in z, i.e., $E = \mathbb{C}(z) \simeq \mathbb{C}(X)$, and the natural map $\mathrm{Aut}(\mathbb{P}_\mathbb{C}^1) \to \mathrm{Aut}(E/\mathbb{C})$ is an isomorphism.

(c) The group $\mathrm{Aut}(\mathbb{C}(X_1, X_2)/\mathbb{C})$ is quite complicated — there is a map

$$\mathrm{PGL}_3(\mathbb{C}) = \mathrm{Aut}(\mathbb{P}_\mathbb{C}^2) \hookrightarrow \mathrm{Aut}(\mathbb{C}(X_1, X_2)/\mathbb{C}),$$

but this is very far from being surjective. When there are even more variables X, the group is not known. The group $\mathrm{Aut}(\mathbb{C}(X_1,\ldots,X_n)/\mathbb{C})$ is the group of birational automorphisms of projective n-space $\mathbb{P}_\mathbb{C}^n$, and is called the **Cremona group.** Its study is part of algebraic geometry (Wikipedia: Cremona group).

In this section, we'll be concerned with the groups $\mathrm{Aut}(E/F)$ when E is a finite extension of F.

PROPOSITION 3.2 *Let E be a splitting field of a separable polynomial f in $F[X]$; then $\mathrm{Aut}(E/F)$ has order $[E:F]$.*

PROOF. As f is separable, it has $\deg f$ distinct roots in E. Therefore Proposition 2.12 shows that the number of F-homomorphisms $E \to E$ is $[E:F]$. Because E is finite over F, all such homomorphisms are isomorphisms. □

EXAMPLE 3.3 We give examples to show that, in the statement of the proposition, is necessary that E be a *splitting* field of a *separable* polynomial.

Consider a simple extension $E = F[\alpha]$, and let f be a polynomial in $F[X]$ having α as a root. If α is the only root of f in E, then $\mathrm{Aut}(E/F) = 1$ by (2.1b). For example, if $\sqrt[3]{2}$ is the real cube root of 2, then $\mathrm{Aut}(\mathbb{Q}[\sqrt[3]{2}]/\mathbb{Q}) = 1$.

Let F be a field of characteristic $p \neq 0$, let a be an element of F that is not a pth power, and let $E = F[\alpha]$, where α is a root of $f = X^p - a$. Then $f = (X - \alpha)^p$ in E, and so E is a splitting field for f, but as f has only one root in E, $\mathrm{Aut}(E/F) = 1$.

When G is a group of automorphisms of a field E, we set

$$E^G = \mathrm{Inv}(G) = \{\alpha \in E \mid \sigma\alpha = \alpha, \text{ all } \sigma \in G\}.$$

It is a subfield of E, called the the **fixed field** of G.

THEOREM 3.4 (ARTIN) *Let G be a finite group of automorphisms of a field E, then*

$$[E:E^G] \leq (G:1).$$

PROOF. Let $F = E^G$, and let $G = \{\sigma_1,\ldots,\sigma_m\}$ with σ_1 the identity map. It suffices to show that every set $\{\alpha_1,\ldots,\alpha_n\}$ of elements of E with $n > m$ is linearly dependent over F. For such a set, consider the system of linear equations

$$\sigma_1(\alpha_1)X_1 + \cdots + \sigma_1(\alpha_n)X_n = 0$$
$$\vdots \qquad\qquad (6)$$
$$\sigma_m(\alpha_1)X_1 + \cdots + \sigma_m(\alpha_n)X_n = 0$$

with coefficients in E. There are m equations and $n > m$ unknowns, and hence there are nontrivial solutions in E. We choose one $(c_1,\ldots,c_n)$ having the fewest possible nonzero elements. After renumbering the α_i, we may suppose that $c_1 \neq 0$, and then, after multiplying by a scalar, that $c_1 \in F$. With these normalizations, we'll show that all $c_i \in F$, and so the first equation

$$\alpha_1 c_1 + \cdots + \alpha_n c_n = 0$$

(recall that σ_1 is the identity map) is a linear relation on the α_i.

If not all c_i are in F, then $\sigma_k(c_i) \neq c_i$ for some $k \neq 1$ and $i \neq 1$. On applying σ_k to the system of linear equations

$$\sigma_1(\alpha_1)c_1 + \cdots + \sigma_1(\alpha_n)c_n = 0$$
$$\vdots$$
$$\sigma_m(\alpha_1)c_1 + \cdots + \sigma_m(\alpha_n)c_n = 0$$

and using that $\{\sigma_k\sigma_1,\ldots,\sigma_k\sigma_m\} = \{\sigma_1,\ldots,\sigma_m\}$, i.e., σ_k merely permutes the σ_i, we find that

$$(c_1,\sigma_k(c_2),\ldots,\sigma_k(c_i),\ldots)$$

is also a solution to the system of equations (6). On subtracting it from the first solution, we obtain a solution $(0,\ldots,c_i - \sigma_k(c_i),\ldots)$, which is nonzero (look at the ith entry), but has more zeros than the first solution (look at the first entry) — contradiction. $\qquad\square$

COROLLARY 3.5 *Let G be a finite group of automorphisms of a field E; then*

$$G = \mathrm{Aut}(E/E^G).$$

PROOF. As $G \subset \mathrm{Aut}(E/E^G)$, we have inequalities

$$[E:E^G] \overset{3.4}{\leq} (G{:}1) \leq (\mathrm{Aut}(E/E^G){:}1) \overset{2.14a}{\leq} [E:E^G].$$

All the inequalities must be equalities, and so $G = \mathrm{Aut}(E/E^G)$. $\qquad\square$

Separable, normal, and Galois extensions

DEFINITION 3.6 An element α algebraic over a field F is *separable* over F if its minimal polynomial over F is separable. An algebraic extension E/F is *separable* if every element of E is separable over F; otherwise, it is *inseparable*.

Thus, an algebraic extension E/F is separable if every irreducible polynomial in $F[X]$ having a root in E is separable, and it is inseparable if

- $\diamond$ F is nonperfect, and in particular has characteristic $p \neq 0$, *and*

- $\diamond$ there is an element α of E whose minimal polynomial is of the form $g(X^p)$, $g \in F[X]$.

See 2.22 *et seq.* For example, the extension $\mathbb{F}_p(T)$ of $\mathbb{F}_p(T^p)$ is inseparable because T has minimal polynomial $X^p - T^p$.

DEFINITION 3.7 An algebraic extension E/F is *normal* if it is algebraic and the minimal polynomial of every element of E splits in $E[X]$.

Thus, an algebraic extension E/F is normal if every irreducible polynomial in $F[X]$ having at least one root in E splits in $E[X]$.

Let E be an algebraic extension of F, and let f be a monic irreducible polynomial in $F[X]$. If f has a root in E, so that it is the minimal polynomial of an element of E, then

$$\left.\begin{array}{ll} E/F \text{ normal} & \Longrightarrow f \text{ splits in } E \\ E/F \text{ separable} & \Longrightarrow f \text{ has only simple roots} \end{array}\right\} \Longrightarrow \begin{array}{l} f \text{ has deg } f \\ \text{distinct roots in } E. \end{array}$$

It follows that E/F is normal and separable if and only if every irreducible polynomial in $F[X]$ having a root in E has $\deg(f)$ distinct roots in E.

EXAMPLE 3.8 (a) The polynomial $X^3 - 2$ has one real root $\sqrt[3]{2}$ and two nonreal roots in $\mathbb{C}$. Therefore the extension $\mathbb{Q}[\sqrt[3]{2}]/\mathbb{Q}$ (which is separable) is not normal.

(b) The extension $\mathbb{F}_p(T)/\mathbb{F}_p(T^p)$ (which is normal) is not separable because the minimal polynomial of T is not separable.

DEFINITION 3.9 An extension E/F of fields is *Galois* if it is finite, normal, and separable. In this case, $\mathrm{Aut}(E/F)$ is called the *Galois group* of E over F, and denoted by $\mathrm{Gal}(E/F)$.

THEOREM 3.10 *For an extension E/F, the following statements are equivalent:*

(a) *E is the splitting field of a separable polynomial $f \in F[X]$;*

(b) *E is finite over F and $F = E^{\mathrm{Aut}(E/F)}$;*

(c) *$F = E^G$ for some finite group G of automorphisms of E;*

(d) *E is Galois over F.*

PROOF. (a) $\Rightarrow$ (b). Certainly, E is finite over F. Let $F' = E^{\mathrm{Aut}(E/F)} \supset F$. We have to show that $F' = F$. Note that E is also the splitting field of f regarded as a polynomial with coefficients in F', and that f is still separable when it is regarded in this way. Hence

$$\left| \mathrm{Aut}(E/F') \right| \overset{3.2}{=} [E:F'] \leq [E:F] \overset{3.2}{=} \left| \mathrm{Aut}(E/F) \right|.$$

According to Corollary 3.5, $\mathrm{Aut}(E/F) = \mathrm{Aut}(E/F')$, and so $[E:F'] = [E:F]$ and $F' = F$.

(b) $\Rightarrow$ (c). Let $G = \mathrm{Aut}(E/F)$. We are given that $F = E^G$, and G is finite because E is finite over F (apply 2.14a).

(c) $\Rightarrow$ (d). According to Theorem 3.4, $[E:F] \leq (G:1)$; in particular, E/F is finite. Let $\alpha \in E$, and let f be the minimal polynomial of α; we have to show that f splits into distinct factors in $E[X]$. Let $\{\alpha_1 = \alpha, \alpha_2, ..., \alpha_m\}$ be the orbit of α under the action of G on E (so the α_i are distinct elements of E), and let

$$g(X) = \prod\nolimits_{i=1}^{m} (X - \alpha_i) = X^m + a_1 X^{m-1} + \cdots + a_m.$$

The coefficients a_j are symmetric polynomials in the α_i, and each $\sigma \in G$ permutes the α_i, and so $\sigma a_j = a_j$ for all j. Thus $g(X) \in F[X]$. As it is monic and $g(\alpha) = 0$, it is divisible by f (see the definition of minimal polynomial, p. 16). Let $\alpha_i = \sigma \alpha$; on applying σ to the equation $f(\alpha) = 0$ we find that $f(\alpha_i) = 0$. Therefore every α_i is a root of f, and so g divides f. Hence $f = g$, and we conclude that $f(X)$ splits into distinct factors in E.

(d) $\Rightarrow$ (a). Because E has finite degree over F, it is generated over F by a finite number of elements, say, $E = F[\alpha_1, ..., \alpha_m]$, $\alpha_i \in E$, α_i algebraic over F. Let f_i be the minimal polynomial of α_i over F, and let f be the product of the distinct f_i. Because E is normal over F, each f_i splits in E, and so E is the splitting field of f. Because E is separable over F, each f_i is separable, and so f is separable. $\qquad\qquad\square$

COROLLARY 3.11 (ARTIN'S THEOREM) *Let G be a finite group of automorphisms of a field E, and let $F = E^G$. Then E is a Galois extension of F with Galois group G, and $[E:F] = (G:1)$.*

PROOF. That E is Galois over F follows from the theorem; that $\mathrm{Gal}(E/F) = G$ follows from 3.5; that $[E:F] = |\mathrm{Gal}(E/F)|$ follows from 3.2.   □

COROLLARY 3.12 *Every finite separable extension E of F is contained in a Galois extension.*

PROOF. Let $E = F[\alpha_1, ..., \alpha_m]$, and let f_i be the minimal polynomial of α_i over F. The product of the distinct f_i is a separable polynomial in $F[X]$ whose splitting field is a Galois extension of F containing E.   □

COROLLARY 3.13 *Let $E \supset M \supset F$; if E is Galois over F, then it is Galois over M.*

PROOF. We know E is the splitting field of some separable $f \in F[X]$; it is also the splitting field of f regarded as an element of $M[X]$.   □

REMARK 3.14 Let E be Galois over F with Galois group G, and let $\alpha \in E$. The elements $\alpha_1, \alpha_2, ..., \alpha_m$ of the orbit of α under G are called the ***conjugates*** of α. In the course of proving the theorem we showed that the minimal polynomial of α is $\prod(X - \alpha_i)$, i.e., the conjugates of α are exactly the roots of its minimal polynomial in E.

REMARK 3.15 Recall that an element α of an algebraic extension of F is said to be separable over F if its minimal polynomial over F is separable. The proof of Corollary 3.12 shows that every finite extension generated by separable elements is separable. Therefore, the elements of an algebraic extension E of F that are separable over F form a subfield E_{sep} of E that is separable over F. This is called the ***separable closure*** of F in E. When E is finite over F, we let $[E:F]_{\mathrm{sep}} = [E_{\mathrm{sep}}:F]$ and call it the ***separable degree*** of E over F.

An algebraic extension E is ***purely inseparable*** over F if the only elements of E separable over F are the elements of F. If E is a finite extension of F, then E is purely inseparable over E_{sep}. See Jacobson, *Lectures in Abstract Algebra*, 1964, Vol. III, Chap. I, Section 10, for more on this topic.

DEFINITION 3.16 An extension E of F is ***cyclic*** (resp. ***abelian***, resp. ***solvable***, etc.) if it is Galois with cyclic (resp. abelian, resp. solvable, etc.) Galois group.

The fundamental theorem of Galois theory

Let E be an extension of F. A ***subextension*** of E/F is an extension M/F with $M \subset E$, i.e., a field M with $F \subset M \subset E$. When E is Galois over F, the subextensions of E/F are in one-to-one correspondence with the subgroups of $\mathrm{Gal}(E/F)$. More precisely, there is the following statement.

THEOREM 3.17 (FUNDAMENTAL THEOREM OF GALOIS THEORY) *Let E be a Galois extension of F with Galois group G. The map $H \mapsto E^H$ is a bijection from the set of subgroups of G to the set of subextensions of E/F,*

$$\{\text{subgroups } H \text{ of } G\} \overset{1:1}{\longleftrightarrow} \{\text{subextensions } F \subset M \subset E\},$$

with inverse $M \mapsto \mathrm{Gal}(E/M)$. Moreover,

(a) $H_1 \supset H_2 \iff E^{H_1} \subset E^{H_2}$ *(the correspondence is order reversing);*

(b) $(H_1 : H_2) = [E^{H_2} : E^{H_1}];$

(c) $\sigma H \sigma^{-1} \leftrightarrow \sigma M$, *i.e.,*

$$E^{\sigma H \sigma^{-1}} = \sigma(E^H);$$

$$\mathrm{Gal}(E/\sigma M) = \sigma \, \mathrm{Gal}(E/M)\sigma^{-1};$$

(d) *H is normal in $G \iff E^H$ is normal (hence Galois) over F, in which case*

$$\mathrm{Gal}(E^H/F) \simeq G/H.$$

PROOF. For the first statement, we have to show that $H \mapsto E^H$ and $M \mapsto \mathrm{Gal}(E/M)$ are inverse maps. Let H be a subgroup of G. Then, Corollary 3.11 shows that $\mathrm{Gal}(E/E^H) = H$. Let M/F be a subextension. Then E is Galois over M by 3.13, which means that $E^{\mathrm{Gal}(E/M)} = M$.

(a) We have the obvious implications,

$$H_1 \supset H_2 \implies E^{H_1} \subset E^{H_2} \implies \mathrm{Gal}(E/E^{H_1}) \supset \mathrm{Gal}(E/E^{H_2}).$$

As $\mathrm{Gal}(E/E^{H_i}) = H_i$, this proves (a).

(b) Let H be a subgroup of G. According to 3.11,

$$(\mathrm{Gal}(E/E^H) : 1) = [E : E^H].$$

This proves (b) in the case $H_2 = 1$, and the general case follows, using that

$$(H_1 : 1) = (H_1 : H_2)(H_2 : 1)$$

$$[E : E^{H_1}] \overset{1.20}{=} [E : E^{H_2}][E^{H_2} : E^{H_1}].$$

(c) For $\tau \in G$ and $\alpha \in E$,

$$\tau\alpha = \alpha \iff \sigma\tau\sigma^{-1}(\sigma\alpha) = \sigma\alpha.$$

Therefore, τ fixes M if and only if $\sigma\tau\sigma^{-1}$ fixes σM, and so $\mathrm{Gal}(E/\sigma M) = \sigma\,\mathrm{Gal}(E/M)\sigma^{-1}$. This shows that $\sigma\,\mathrm{Gal}(E/M)\sigma^{-1}$ corresponds to σM.

(d) Let H be a normal subgroup of G. Because $\sigma H\sigma^{-1} = H$ for all $\sigma \in G$, we must have $\sigma E^H = E^H$ for all $\sigma \in G$, i.e., the action of G on E stabilizes E^H. We therefore have a homomorphism

$$\sigma \mapsto \sigma|E^H : G \to \mathrm{Aut}(E^H/F)$$

whose kernel is H. As $(E^H)^{G/H} = F$, we see that E^H is Galois over F (by Theorem 3.10) and that $G/H \simeq \mathrm{Gal}(E^H/F)$ (by 3.11).

Conversely, suppose that M is normal over F, and let $\alpha_1,\ldots,\alpha_m$ generate M over F. For $\sigma \in G$, $\sigma\alpha_i$ is a root of the minimal polynomial of α_i over F, and so lies in M. Hence $\sigma M = M$, and this implies that $\sigma H\sigma^{-1} = H$ (by (c)). $\qquad\square$

REMARK 3.18 Let E/F be a Galois extension, so that there is an order reversing bijection between the subextensions of E/F and the subgroups of G. From this, we can read off the following results.

(a) Let $M_1, M_2, \ldots, M_r$ be subextensions of E/F, and let H_i be the subgroup corresponding to M_i (i.e., $H_i = \mathrm{Gal}(E/M_i)$). Then (by definition) $M_1 M_2 \cdots M_r$ is the smallest field containing all M_i; hence it must correspond to the largest subgroup contained in all H_i, which is $\bigcap H_i$. We have shown that
$$\mathrm{Gal}(E/M_1 \cdots M_r) = H_1 \cap \ldots \cap H_r.$$

(b) Let H be a subgroup of G and let $M = E^H$. The largest normal subgroup contained in H is $N = \bigcap_{\sigma \in G} \sigma H\sigma^{-1}$ (see GT, 4.10), and so E^N is the smallest normal extension of F containing M. Note that, by (a), E^N is the composite of the fields σM. It is called the **normal**, or **Galois**, closure of M in E.

PROPOSITION 3.19 *Let E and L be extensions of F contained in some common field. If E/F is Galois, then EL/L and $E/E\cap L$ are Galois, and the map*
$$\sigma \mapsto \sigma|E : \mathrm{Gal}(EL/L) \to \mathrm{Gal}(E/E\cap L)$$

is an isomorphism.

PROOF. Because E is Galois over F, it is the splitting field of a separable polynomial $f \in F[X]$. Then EL is the splitting field of f over L, and E is the splitting field of f over $E \cap L$. Hence EL/L and $E/E \cap L$ are Galois. Every automorphism σ of EL fixing the elements of L maps roots of f to roots of f, and so $\sigma E = E$. There is therefore a homomorphism

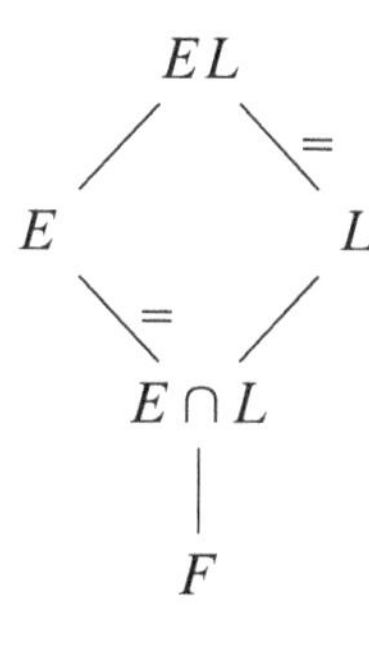

$$\sigma \mapsto \sigma|E \colon \mathrm{Gal}(EL/L) \to \mathrm{Gal}(E/E \cap L).$$

If $\sigma \in \mathrm{Gal}(EL/L)$ fixes the elements of E, then it fixes the elements of EL, and hence is the identity map. Thus, $\sigma \mapsto \sigma|E$ is injective. If $\alpha \in E$ is fixed by all $\sigma \in \mathrm{Gal}(EL/L)$, then $\alpha \in E \cap L$. By Corollary 3.5, this implies that the image of $\sigma \mapsto \sigma|E$ is $\mathrm{Gal}(E/E \cap L)$. $\square$

COROLLARY 3.20 *Suppose, in the proposition, that L is finite over F. Then*

$$[EL \colon F] = \frac{[E \colon F][L \colon F]}{[E \cap L \colon F]}.$$

PROOF. According to Proposition 1.20,

$$[EL \colon F] = [EL \colon L][L \colon F],$$

but

$$[EL \colon L] \overset{3.19}{=} [E \colon E \cap L] \overset{1.20}{=} \frac{[E \colon F]}{[E \cap L \colon F]}.$$ $\square$

PROPOSITION 3.21 *Let E_1 and E_2 be extensions of F contained in some common field. If E_1 and E_2 are Galois over F, then $E_1 E_2$ and $E_1 \cap E_2$ are Galois over F, and the map*

$$\sigma \mapsto (\sigma|E_1, \sigma|E_2) \colon \mathrm{Gal}(E_1 E_2/F) \to \mathrm{Gal}(E_1/F) \times \mathrm{Gal}(E_2/F)$$

is an isomorphism of $\mathrm{Gal}(E_1 E_2/F)$ onto the subgroup

$$H = \{(\sigma_1, \sigma_2) \mid \sigma_1|E_1 \cap E_2 = \sigma_2|E_1 \cap E_2\}$$

of $\mathrm{Gal}(E_1/F) \times \mathrm{Gal}(E_2/F)$.

In other words,

$$\mathrm{Gal}(E_1 E_2/F) \simeq \mathrm{Gal}(E_1/F) \underset{\mathrm{Gal}(E_1 \cap E_2/F)}{\times} \mathrm{Gal}(E_2/F).$$

PROOF: Let $a \in E_1 \cap E_2$, and let f be its minimal polynomial over F. Then f has $\deg f$ distinct roots in E_1 and $\deg f$ distinct roots in E_2. Since f can have at most $\deg f$ roots in $E_1 E_2$, it follows that it has $\deg f$ distinct roots in $E_1 \cap E_2$. This shows that $E_1 \cap E_2$ is normal and separable over F, and hence Galois (3.10). As E_1 and E_2 are Galois over F, they are splitting fields for separable polynomials $f_1, f_2 \in F[X]$. Now $E_1 E_2$ is a splitting field for $\mathrm{lcm}(f_1, f_2)$, and hence it also is Galois over F. The map $\sigma \mapsto (\sigma|E_1, \sigma|E_2)$ is clearly an injective homomorphism, and its image is contained in H. We'll prove that the image is the whole of H by counting.

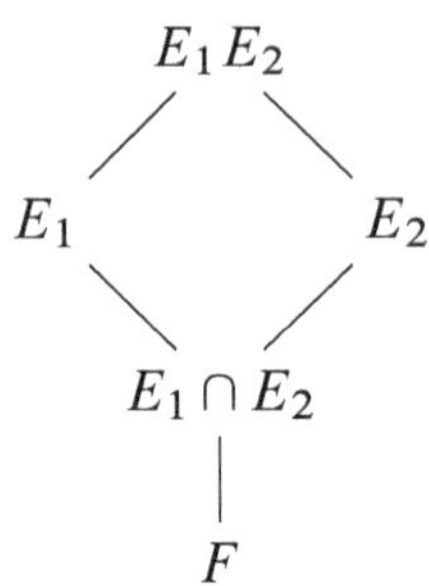

From the fundamental theorem,

$$\frac{\mathrm{Gal}(E_2/F)}{\mathrm{Gal}(E_2/E_1 \cap E_2)} \simeq \mathrm{Gal}(E_1 \cap E_2/F),$$

and so, for each $\sigma_1 \in \mathrm{Gal}(E_1/F)$, $\sigma_1|E_1 \cap E_2$ has exactly $[E_2 : E_1 \cap E_2]$ extensions to an element of $\mathrm{Gal}(E_2/F)$. Therefore,

$$(H:1) = [E_1:F][E_2:E_1 \cap E_2] = \frac{[E_1:F] \cdot [E_2:F]}{[E_1 \cap E_2:F]},$$

which equals $[E_1 E_2 : F]$ by (3.20). $\square$

Examples

EXAMPLE 3.22 We analyse the extension $\mathbb{Q}[\zeta]/\mathbb{Q}$, where ζ is a primitive 7th root of 1, say $\zeta = e^{2\pi i/7}$.

Note that $\mathbb{Q}[\zeta]$ is the splitting field of the polynomial $X^7 - 1$, and that ζ has minimal polynomial

$$X^6 + X^5 + X^4 + X^3 + X^2 + X + 1$$

(see 1.42). Therefore, $\mathbb{Q}[\zeta]$ is Galois of degree 6 over $\mathbb{Q}$. For any $\sigma \in \mathrm{Gal}(\mathbb{Q}[\zeta]/\mathbb{Q})$, $\sigma\zeta = \zeta^i$, some i, $1 \le i \le 6$, and the map $\sigma \mapsto i$ defines an isomorphism $\mathrm{Gal}(\mathbb{Q}[\zeta]/\mathbb{Q}) \to (\mathbb{Z}/7\mathbb{Z})^\times$. Let σ be the element of $\mathrm{Gal}(\mathbb{Q}[\zeta]/\mathbb{Q})$ such that $\sigma\zeta = \zeta^3$. Then σ generates $\mathrm{Gal}(\mathbb{Q}[\zeta]/\mathbb{Q})$ because the class of 3 in $(\mathbb{Z}/7\mathbb{Z})^\times$ generates it (the powers of 3 mod 7 are $3, 2, 6, 4, 5, 1$). We investigate the subfields of $\mathbb{Q}[\zeta]$ corresponding to the subgroups $\langle \sigma^3 \rangle$ and $\langle \sigma^2 \rangle$.

Note that $\sigma^3\zeta = \zeta^6 = \bar{\zeta}$ (complex conjugate of ζ), and so $\zeta + \bar{\zeta} = 2\cos\frac{2\pi}{7}$ is fixed by σ^3. Now $\mathbb{Q}[\zeta] \supset \mathbb{Q}[\zeta]^{\langle\sigma^3\rangle} \supset \mathbb{Q}[\zeta + \bar{\zeta}] \neq \mathbb{Q}$, and so $\mathbb{Q}[\zeta]^{\langle\sigma^3\rangle} = \mathbb{Q}[\zeta + \bar{\zeta}]$ (look at degrees). As $\langle\sigma^3\rangle$ is a normal subgroup of $\langle\sigma\rangle$, $\mathbb{Q}[\zeta + \bar{\zeta}]$ is Galois over $\mathbb{Q}$, with Galois group $\langle\sigma\rangle/\langle\sigma^3\rangle$. The conjugates of $\alpha_1 \overset{\text{def}}{=} \zeta + \bar{\zeta}$ are $\alpha_3 = \zeta^3 + \zeta^{-3}$, $\alpha_2 = \zeta^2 + \zeta^{-2}$. Direct calculation shows that

$$\alpha_1 + \alpha_2 + \alpha_3 = \sum_{i=1}^{6} \zeta^i = -1,$$

$$\alpha_1\alpha_2 + \alpha_1\alpha_3 + \alpha_2\alpha_3 = -2,$$

$$\begin{aligned}
\alpha_1\alpha_2\alpha_3 &= (\zeta + \zeta^6)(\zeta^2 + \zeta^5)(\zeta^3 + \zeta^4) \\
&= (\zeta + \zeta^3 + \zeta^4 + \zeta^6)(\zeta^3 + \zeta^4) \\
&= (\zeta^4 + \zeta^6 + 1 + \zeta^2 + \zeta^5 + 1 + \zeta + \zeta^3) \\
&= 1.
\end{aligned}$$

Hence the minimal polynomial[1] of $\zeta + \bar{\zeta}$ is

$$g(X) = X^3 + X^2 - 2X - 1.$$

The minimal polynomial of $\cos\frac{2\pi}{7} = \frac{\alpha_1}{2}$ is therefore

$$\frac{g(2X)}{8} = X^3 + X^2/2 - X/2 - 1/8.$$

The subfield of $\mathbb{Q}[\zeta]$ corresponding to $\langle\sigma^2\rangle$ is generated by $\beta = \zeta + \zeta^2 + \zeta^4$. Let $\beta' = \sigma\beta$. Then $(\beta - \beta')^2 = -7$. Hence the field fixed by $\langle\sigma^2\rangle$ is $\mathbb{Q}[\sqrt{-7}]$.

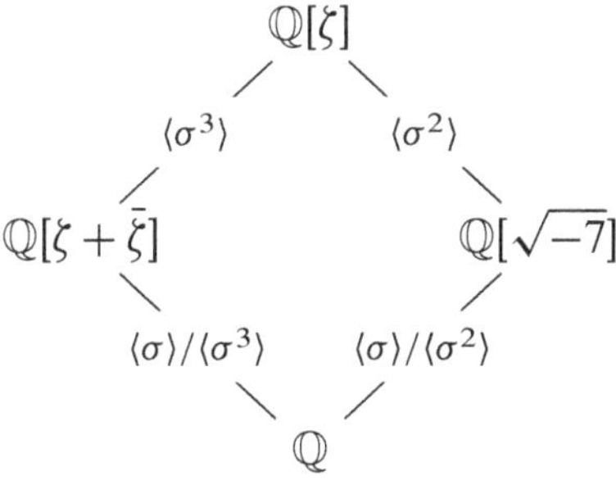

---

[1]More directly, on setting $X = \zeta + \bar{\zeta}$ in

$$(X^3 - 3X) + (X^2 - 2) + X + 1$$

one obtains $1 + \zeta + \zeta^2 + \cdots + \zeta^6 = 0$.

EXAMPLE 3.23 We compute the Galois group of a splitting field E of $X^5 - 2 \in \mathbb{Q}[X]$ (see the diagram on the cover).

Recall from Exercise 2-3 that $E = \mathbb{Q}[\zeta, \alpha]$ where ζ is a primitive 5th root of 1, and α is a root of $X^5 - 2$. For example, we could take E to be the splitting field of $X^5 - 2$ in $\mathbb{C}$, with $\zeta = e^{2\pi i/5}$ and α equal to the real 5th root of 2. We have the picture at right, and

$$[\mathbb{Q}[\zeta] : \mathbb{Q}] = 4, \quad [\mathbb{Q}[\alpha] : \mathbb{Q}] = 5.$$

Because 4 and 5 are relatively prime,

$$[\mathbb{Q}[\zeta, \alpha] : \mathbb{Q}] = 20.$$

Hence $G = \mathrm{Gal}(\mathbb{Q}[\zeta, \alpha]/\mathbb{Q})$ has order 20, and the subgroups N and H fixing $\mathbb{Q}[\zeta]$ and $\mathbb{Q}[\alpha]$ have orders 5 and 4 respectively. Because $\mathbb{Q}[\zeta]$ is normal over $\mathbb{Q}$ (it is the splitting field of $X^5 - 1$), N is normal in G. Because $\mathbb{Q}[\zeta] \cdot \mathbb{Q}[\alpha] = \mathbb{Q}[\zeta, \alpha]$, we have $H \cap N = 1$, and so $G = N \rtimes_\theta H$. Moreover, $H \simeq G/N \simeq (\mathbb{Z}/5\mathbb{Z})^\times$, which is cyclic, being generated by the class of 2. Let τ be the generator of H corresponding to 2 under this isomorphism, and let σ be a generator of N. Thus $\sigma(\alpha)$ is another root of $X^5 - 2$, which we can take to be $\zeta\alpha$ (after possibly replacing σ by a power). Hence:

$$\begin{cases} \tau\zeta &= \zeta^2 \\ \tau\alpha &= \alpha \end{cases} \qquad \begin{cases} \sigma\zeta &= \zeta \\ \sigma\alpha &= \zeta\alpha. \end{cases}$$

Note that $\tau\sigma\tau^{-1}(\alpha) = \tau\sigma\alpha = \tau(\zeta\alpha) = \zeta^2\alpha$ and it fixes ζ; therefore $\tau\sigma\tau^{-1} = \sigma^2$. Thus G has generators σ and τ and defining relations

$$\sigma^5 = 1, \quad \tau^4 = 1, \quad \tau\sigma\tau^{-1} = \sigma^2.$$

The subgroup H has five conjugates, which correspond to the five fields $\mathbb{Q}[\zeta^i\alpha]$,

$$\sigma^i H \sigma^{-i} \leftrightarrow \sigma^i \mathbb{Q}[\alpha] = \mathbb{Q}[\zeta^i\alpha], \qquad 1 \le i \le 5.$$

Constructible numbers revisited

Earlier (1.37) we showed that a real number α is constructible if and only if it is contained in a subfield of $\mathbb{R}$ of the form $\mathbb{Q}[\sqrt{a_1}, \ldots, \sqrt{a_r}]$ with each a_i a positive element of $\mathbb{Q}[\sqrt{a_1}, \ldots, \sqrt{a_{i-1}}]$. In particular

$$\alpha \text{ constructible} \implies [\mathbb{Q}[\alpha] : \mathbb{Q}] = 2^s \text{ some } s. \tag{7}$$

Now we can prove a partial converse to this last statement.

THEOREM 3.24 *If α is contained in a subfield of $\mathbb{R}$ that is Galois of degree 2^r over $\mathbb{Q}$, then it is constructible.*

PROOF. Suppose $\alpha \in E \subset \mathbb{R}$ where E is Galois of degree 2^r over $\mathbb{Q}$, and let $G = \mathrm{Gal}(E/\mathbb{Q})$. Because finite p-groups are solvable (GT, 6.7), there exists a sequence of groups

$$\{1\} = G_0 \subset G_1 \subset G_2 \subset \cdots \subset G_r = G$$

with G_i/G_{i-1} of order 2. Correspondingly, there will be a sequence of fields,

$$E = E_0 \supset E_1 \supset E_2 \supset \cdots \supset E_r = \mathbb{Q}$$

with E_{i-1} of degree 2 over E_i. The next lemma shows that $E_i = E_{i-1}[\sqrt{a_i}]$ for some $a_i \in E_{i-1}$, and $a_i > 0$ because otherwise E_i would not be real. This proves the theorem. $\square$

LEMMA 3.25 *Let E/F be a quadratic extension of fields of characteristic $\neq 2$. Then $E = F[\sqrt{d}]$ for some $d \in F$.*

PROOF. Let $\alpha \in E$, $\alpha \notin F$, and let $X^2 + bX + c$ be the minimal polynomial of α. Then $\alpha = \frac{-b \pm \sqrt{b^2 - 4c}}{2}$, and so $E = F[\sqrt{b^2 - 4c}]$. $\square$

COROLLARY 3.26 *If p is a prime of the form $2^k + 1$, then $\cos \frac{2\pi}{p}$ is constructible.*

PROOF. The field $\mathbb{Q}[e^{2\pi i/p}]$ is Galois over $\mathbb{Q}$ with Galois group G canonically isomorphic to $(\mathbb{Z}/p\mathbb{Z})^\times$, which has order $p - 1 = 2^k$. The field $\mathbb{Q}[\cos \frac{2\pi}{p}]$ is contained in $\mathbb{Q}[e^{2\pi i/p}]$, and therefore is Galois of degree dividing 2^k (fundamental theorem 3.17 and 1.20). As $\mathbb{Q}[\cos \frac{2\pi}{p}]$ is a subfield of $\mathbb{R}$, we can apply the theorem. $\square$

Thus a regular p-gon, p prime, is constructible if and only if p is a Fermat prime, i.e., of the form $2^{2^r} + 1$. For example, we have proved that the regular 65537-polygon is constructible, without (happily) having to exhibit an explicit formula for $\cos \frac{2\pi}{65537}$.

REMARK 3.27 The converse to (7) is false. In fact, there are nonconstructible algebraic numbers of degree 4 over $\mathbb{Q}$.

For example, the polynomial $f(X) = X^4 - 4X + 2 \in \mathbb{Q}[X]$ is irreducible, and we show below (4.10) that the Galois group of its splitting field E is S_4. If the four roots of f were constructible, then every element of E would be constructible (1.36), but S_4 has a subgroup H of order 8, and E^H has degree 3 over $\mathbb{Q}$, and so no element of $E^H \smallsetminus \mathbb{Q}$ is constructible.

Alternatively, if a root α of $f(X)$ were constructible, then there would exist a tower of quadratic extensions $\mathbb{Q}[\alpha] \supset M \supset \mathbb{Q}$. By Galois theory, the groups $\mathrm{Gal}(E/M) \supset \mathrm{Gal}(E/\mathbb{Q}[\alpha])$ have orders 12 and 6 respectively. As $\mathrm{Gal}(E/\mathbb{Q}) = S_4$, $\mathrm{Gal}(E/M)$ would be A_4. But A_4 has no subgroup of order 6, a contradiction.

The Galois group of a polynomial

If a polynomial $f \in F[X]$ is separable, then its splitting field F_f is Galois over F, and we call $\mathrm{Gal}(F_f/F)$ the **Galois group** G_f of f.

Let $f(X) = \prod_{i=1}^{n}(X - \alpha_i)$ in a splitting field F_f. We know that the elements of $\mathrm{Gal}(F_f/F)$ map roots of f to roots of f, i.e., they map the set $\{\alpha_1, \alpha_2, \ldots, \alpha_n\}$ into itself. Being automorphisms, they act as permutations on $\{\alpha_1, \alpha_2, \ldots, \alpha_n\}$. As the α_i generate F_f over F, an element of $\mathrm{Gal}(F_f/F)$ is uniquely determined by the permutation it defines. Thus G_f can be identified with a subset of $\mathrm{Sym}(\{\alpha_1, \alpha_2, \ldots, \alpha_n\}) \approx S_n$ (symmetric group on n symbols). In fact, G_f consists exactly of the permutations σ of $\{\alpha_1, \alpha_2, \ldots, \alpha_n\}$ such that, for $P \in F[X_1, \ldots, X_n]$,

$$P(\alpha_1, \ldots, \alpha_n) = 0 \implies P(\sigma\alpha_1, \ldots, \sigma\alpha_n) = 0. \tag{8}$$

To see this, note that the kernel of the map

$$F[X_1, \ldots, X_n] \to F_f, \quad X_i \mapsto \alpha_i, \tag{9}$$

consists of the polynomials $P(X_1, \ldots, X_n)$ such that $P(\alpha_1, \ldots, \alpha_n) = 0$. Let σ be a permutation of the α_i satisfying the condition (8). Then the map

$$F[X_1, \ldots, X_n] \to F_f, \quad X_i \mapsto \sigma\alpha_i,$$

factors through the map (9), and defines an F-isomorphism $F_f \to F_f$, i.e., an element of the Galois group. This shows that every permutation satisfying the condition (8) extends uniquely to an element of G_f, and it is obvious that every element of G_f arises in this way.

This gives a description of G_f not mentioning fields or abstract groups, neither of which were available to Galois. Note that it shows again that $(G_f:1)$, hence $[F_f:F]$, divides $\deg(f)!$.

Solvability of equations

For a polynomial $f \in F[X]$, we say that $f(X) = 0$ is ***solvable in radicals*** if its solutions can be obtained by the algebraic operations of addition, subtraction, multiplication, division, and the extraction of mth roots, or, more precisely, if there exists a tower of fields

$$F = F_0 \subset F_1 \subset F_2 \subset \cdots \subset F_m$$

such that

   (a) $F_i = F_{i-1}[\alpha_i]$, $\alpha_i^{m_i} \in F_{i-1}$;

   (b) F_m contains a splitting field for f.

THEOREM 3.28 (GALOIS, 1832) *Let F be a field of characteristic zero, and let $f \in F[X]$. The equation $f(X) = 0$ is solvable in radicals if and only if the Galois group of f is solvable.*

We'll prove this later (5.34). Also we'll exhibit polynomials $f(X) \in \mathbb{Q}[X]$ with Galois group S_n, which are therefore not solvable when $n \geq 5$ by GT, 4.37.

REMARK 3.29 When F has characteristic p, the theorem fails for two reasons,

   (a) f need not be separable, and so not have a Galois group;

   (b) $X^p - X - a = 0$ need not be solvable in radicals even though it is separable with abelian Galois group (cf. Exercise 2-2).

If the definition of solvable is changed to allow extensions defined by polynomials of the type in (b) in the chain, then the theorem holds for fields F of characteristic $p \neq 0$ and separable $f \in F[X]$.

ASIDE 3.30 Abel (1828) proved the following statement: If three roots of an arbitrary irreducible equation of a prime degree are such that one of them can be rationally expressed as function of the other two, then the equation can be solved by radicals. Sylow (1902) claimed that Abel's statement is incorrect. Deligne (C. R. Math. Acad. Sci. Paris 359 (2021), 919–921) supplied a complete and elegant proof for Abel's statement (including its converse) by using the Galois theory of fields to convert everything into the language of permutation groups. He proves that if E is a set with p elements (p a prime number) and G is a transitive group of permutations of E, then G is a solvable group if and only if, for any three elements of E, there exists one of them that is fixed by any element g of G such that g fixes the other two elements.

NOTES Much of what has been written about Galois is unreliable — see Tony Rothman, *Genius and Biographers: The Fictionalization of Evariste Galois*, Amer. Math. Monthly, 89, 84 (1982). For a careful explanation of Galois's "Premier Mémoire", see Harold Edwards, *Galois for 21st-century readers*. Notices Amer. Math. Soc. 59 (2012), no. 7, 912–923.

Exercises

3-1 Let F be a field of characteristic 0. Show that $F(X^2) \cap F(X^2 - X) = F$ (intersection inside $F(X)$). [Hint: Find automorphisms σ and τ of $F(X)$, each of order 2, fixing $F(X^2)$ and $F(X^2 - X)$ respectively, and show that $\sigma\tau$ has infinite order.]

3-2 Let p be an odd prime, and let ζ be a primitive pth root of 1 in $\mathbb{C}$. Let $E = \mathbb{Q}[\zeta]$, and let $G = \mathrm{Gal}(E/\mathbb{Q})$; thus $G = (\mathbb{Z}/(p))^\times$. Let H be the subgroup of index 2 in G. Put $\alpha = \sum_{i \in H} \zeta^i$ and $\beta = \sum_{i \in G \setminus H} \zeta^i$. Show:

(a) α and β are fixed by H;

(b) if $\sigma \in G \setminus H$, then $\sigma\alpha = \beta$, $\sigma\beta = \alpha$.

Thus α and β are roots of the polynomial $X^2 + X + \alpha\beta \in \mathbb{Q}[X]$. Compute $\alpha\beta$ (or $\alpha - \beta$) and show that the fixed field of H is $\mathbb{Q}[\sqrt{p}]$ when $p \equiv 1 \bmod 4$ and $\mathbb{Q}[\sqrt{-p}]$ when $p \equiv 3 \bmod 4$.[2]

3-3 Let $M = \mathbb{Q}[\sqrt{2}, \sqrt{3}]$ and $E = M[\sqrt{(\sqrt{2}+2)(\sqrt{3}+3)}]$ (subfields of $\mathbb{R}$).

(a) Show that M is Galois over $\mathbb{Q}$ with Galois group the 4-group $C_2 \times C_2$.

(b) Show that E is Galois over $\mathbb{Q}$ with Galois group the quaternion group.

3-4 Let E be a Galois extension of F with Galois group G, and let L be the fixed field of a subgroup H of G. Show that the automomorphism group of L/F is N/H where N is the normalizer of H in G.

3-5 Let E be a finite extension of F. Show that the order of $\mathrm{Aut}(E/F)$ divides the degree $[E:F]$.

---

[2]This problem shows that every quadratic extension of $\mathbb{Q}$ is contained in a cyclotomic extension of $\mathbb{Q}$. The Kronecker-Weber theorem says that *every* abelian extension of $\mathbb{Q}$ is contained in a cyclotomic extension.

CHAPTER 4

Computing Galois Groups

In this chapter, we investigate general methods for computing Galois groups.

When is $G_f \subset A_n$?

Let σ be a permutation of the set $\{1, 2, \ldots, n\}$. The pairs (i, j) with $i < j$ but $\sigma(i) > \sigma(j)$ are called the **inversions** of σ, and σ is said to be **even** or **odd** according as the number of inversions is even or odd. The **signature** of σ, $\mathrm{sign}(\sigma)$, is $+1$ or -1 according as σ is even or odd. We can define the signature of a permutation σ of any set S of n elements by choosing a numbering of the set and identifying σ with a permutation of $\{1, \ldots, n\}$. The group $\mathrm{Sym}(S)$ of permutations of S is generated by transpositions, and sign is the unique homomorphism $\mathrm{Sym}(S) \to \{\pm 1\}$ such that $\mathrm{sign}(\sigma) = -1$ for every transposition. In particular, it is independent of the choice of the numbering. See GT, 4.25.

Now consider a monic polynomial

$$f(X) = X^n + a_1 X^{n-1} + \cdots + a_n$$

and let $f(X) = \prod_{i=1}^{n}(X - \alpha_i)$ in some splitting field. Set

$$\Delta(f) = \prod_{1 \le i < j \le n}(\alpha_i - \alpha_j), \qquad D(f) = \Delta(f)^2 = \prod_{1 \le i < j \le n}(\alpha_i - \alpha_j)^2.$$

The **discriminant** of f is defined to be $D(f)$. Note that $D(f)$ is nonzero if and only if f has only simple roots, i.e., is separable. Let G_f be the Galois group of f, and identify it with a subgroup of $\mathrm{Sym}(\{\alpha_1, \ldots, \alpha_n\})$ (as on p. 54).

PROPOSITION 4.1 *Let $f \in F[X]$ be a separable polynomial, and let $\sigma \in G_f$.*
   (a) $\sigma\Delta(f) = \text{sign}(\sigma)\Delta(f)$.
   (b) $\sigma D(f) = D(f)$.

PROOF. Each inversion of σ introduces a negative sign into $\sigma\Delta(f)$, and so (a) follows from the definition of $\text{sign}(\sigma)$. The equality in (b) is obtained by squaring that in (a). $\qquad\square$

While $\Delta(f)$ depends on the choice of the numbering of the roots of f, $D(f)$ does not.

COROLLARY 4.2 *Let $f(X) \in F[X]$ be separable of degree n. Let F_f be a splitting field for f and let $G_f = \text{Gal}(F_f/F)$.*
   (a) *The discriminant $D(f) \in F$.*
   (b) *Assume that $\text{char}(F) \neq 2$. The subfield of F_f corresponding to $A_n \cap G_f$ is $F[\Delta(f)]$. Hence*

$$G_f \subset A_n \iff \Delta(f) \in F \iff D(f) \text{ is a square in } F.$$

PROOF. (a) The discriminant of f is an element of F_f fixed by $G_f \overset{\text{def}}{=} \text{Gal}(F_f/F)$, and hence lies in F (by the fundamental theorem).
   (b) Because f has simple roots, $\Delta(f) \neq 0$, and so the formula $\sigma\Delta(f) = \text{sign}(\sigma)\Delta(f)$ shows that an element of G_f fixes $\Delta(f)$ if and only if it lies in A_n. Thus, under the Galois correspondence,

$$G_f \cap A_n \leftrightarrow F[\Delta(f)].$$

Hence,
$$G_f \cap A_n = G_f \iff F[\Delta(f)] = F. \qquad\square$$

The roots of $X^2 + bX + c$ are $\dfrac{-b \pm \sqrt{b^2 - 4c}}{2}$ and so

$$\Delta(X^2 + bX + c) = \sqrt{b^2 - 4c} \ (\text{or } -\sqrt{b^2 - 4c}),$$
$$D(X^2 + bX + c) = b^2 - 4c.$$

Similarly,
$$D(X^3 + bX + c) = -4b^3 - 27c^2.$$

By completing the cube, one can put any cubic polynomial in this form (in characteristic $\neq 3$).

Although there is a not a universal formula for the roots of f in terms of its coefficients when the $\deg(f) > 4$, there is for its discriminant. However, the formulas for the discriminant rapidly become very complicated, for example, that for $X^5 + aX^4 + bX^3 + cX^2 + dX + e$ has 59 terms. Fortunately, PARI knows them. For example, typing `poldisc(X^3+a*X^2+b*X+c,X)` returns the discriminant of $X^3 + aX^2 + bX + c$, namely,

$$-4ca^3 + b^2a^2 + 18cba + (-4b^3 - 27c^2).$$

For an efficient way of calculating discriminants using resultants, see the appendix to this chapter.

REMARK 4.3 Suppose $F \subset \mathbb{R}$. Then $D(f)$ will not be a square if it is negative. It is known that the sign of $D(f)$ is $(-1)^s$ where $2s$ is the number of nonreal roots of f in $\mathbb{C}$ (see ANT 2.40). Thus if s is odd, then G_f is not contained in A_n. This can be proved more directly by noting that $\mathrm{sign}\colon G_f \to \{\pm 1\}$ is surjective because complex conjugation acts on the roots as the product of s disjoint transpositions.

The converse is not true: when s is even, G_f is not necessarily contained in A_n.

ASIDE 4.4 When F has characteristic 2, the discriminant is always a square, and so it is not useful for deciding whether G_f is contained in A_n. Instead, we must use the **Berlekamp discriminant**, which for a separable polynomial $f(X) = \prod_{i=1}^{n}(X - \alpha_i)$ is defined to be

$$D = \sum_{i<j} \frac{\alpha_i \alpha_j}{\alpha_i^2 + \alpha_j^2}.$$

The Galois group G_f of f is contained in A_n if and only if there exists a $\delta \in F$ such that $\delta^2 + \delta = D$. See Berlekamp, *An analog to the discriminant over fields of characteristic two*. J. Algebra 38 (1976), no. 2, 315–317.

When does G_f act transitively on the roots?

PROPOSITION 4.5 *Let $f(X) \in F[X]$ be separable. Then $f(X)$ is irreducible if and only if G_f permutes the roots of f transitively.*

PROOF. $\implies$: Let F_f be a splitting field for f. If α and β are two roots of $f(X)$ in F_f, then they both have $f(X)$ as their minimal polynomial (because f is irreducible), and so $F[\alpha]$ and $F[\beta]$ are both stem fields for f. Hence, there is an F-isomorphism

$$F[\alpha] \simeq F[\beta], \qquad \alpha \leftrightarrow \beta.$$

Write $F_f = F[\alpha_1,\alpha_2,...]$ with $\alpha_1 = \alpha$ and $\alpha_2,\alpha_3,...$ the other roots of $f(X)$. Then the F-homomorphism $\alpha \mapsto \beta\colon F[\alpha] \to F_f$ extends (step by step) to an F-homomorphism $F_f \to F_f$ (use 2.4b), which is an F-isomorphism sending α to β.

$\Longleftarrow$: Let $g(X) \in F[X]$ be an irreducible factor of f, and let α be one of its roots. If β is a second root of f, then (by assumption) $\beta = \sigma\alpha$ for some $\sigma \in G_f$. Now, because g has coefficients in F,

$$g(\sigma\alpha) = \sigma g(\alpha) = 0,$$

and so β is also a root of g. Therefore, every root of f is also a root of g, and so $f(X) = g(X)$. $\qquad\square$

Note that if $f(X)$ is irreducible of degree n, then n divides $(G_f\colon 1)$ because $n = [F[\alpha]\colon F]$, which divides $[F_f\colon F] = (G_f\colon 1)$. Thus G_f is a transitive subgroup of S_n whose order is divisible by n.

Polynomials of degree at most three

EXAMPLE 4.6 Let $f(X) \in F[X]$ be a polynomial of degree 2. When F has odd characteristic and f is not a square,

$$f \text{ is irreducible } \iff D(f) \text{ is not a square } \iff G_f = S_2.$$

In characteristic 2, a quadratic polynomial may be irreducible but not separable (e.g., $f(X) = X^2 - a$ for some $a \in F \smallsetminus F^2$) or irreducible and separable but have discriminant a square (e.g., $f(X) = X^2 - X - a$ for suitable a).

EXAMPLE 4.7 Let $f(X) \in F[X]$ be a polynomial of degree 3, and suppose that $\operatorname{char}(F) \neq 3$. We may assume f to be irreducible, for otherwise we are essentially back in the previous case. Then f is separable and G_f is a transitive subgroup of S_3 whose order is divisible by 3. There are only two possibilities: $G_f = A_3$ or S_3 according as $D(f)$ is a square in F or not. Note that A_3 is generated by the cycle (123).

For example, $X^3 - 3X + 1$ is irreducible in $\mathbb{Q}[X]$ (see 1.12). Its discriminant is $-4(-3)^3 - 27 = 81 = 9^2$, and so its Galois group is A_3.

On the other hand, $X^3 + 3X + 1 \in \mathbb{Q}[X]$ is also irreducible (apply 1.11), but its discriminant is -135, and so its Galois group is S_3.

Quartic polynomials

Let $f(X)$ be a separable quartic polynomial. In order to determine G_f we'll exploit the fact that S_4 has

$$V \overset{\text{def}}{=} \{1, (12)(34), (13)(24), (14)(23)\}$$

as a normal subgroup — it is normal because it contains all elements of type $2+2$ (GT, 4.29). Let E be a splitting field of f, and let $f(X) = (X-\alpha_1)(X-\alpha_2)(X-\alpha_3)(X-\alpha_4)$ in E. We identify the Galois group G_f of f with a subgroup of the symmetric group $\mathrm{Sym}(\{\alpha_1,\alpha_2,\alpha_3,\alpha_4\})$. Consider the partially symmetric elements

$$\alpha = \alpha_1\alpha_2 + \alpha_3\alpha_4$$
$$\beta = \alpha_1\alpha_3 + \alpha_2\alpha_4$$
$$\gamma = \alpha_1\alpha_4 + \alpha_2\alpha_3.$$

They are distinct because the α_i are distinct; for example,

$$\alpha - \beta = \alpha_1(\alpha_2 - \alpha_3) + \alpha_4(\alpha_3 - \alpha_2) = (\alpha_1 - \alpha_4)(\alpha_2 - \alpha_3).$$

The group $\mathrm{Sym}(\{\alpha_1,\alpha_2,\alpha_3,\alpha_4\})$ permutes $\{\alpha,\beta,\gamma\}$ transitively. The stabilizer of each of α,β,γ must therefore be a subgroup of index 3 in S_4, and hence has order 8. For example, the stabilizer of β is $\langle(1234),(13)\rangle$. Groups of order 8 in S_4 are Sylow 2-subgroups. There are three of them, all isomorphic to D_4. By the Sylow theorems, V is contained in a Sylow 2-subgroup; in fact, because the Sylow 2-subgroups are conjugate and V is normal, it is contained in all three. It follows that V is the intersection of the three Sylow 2-subgroups. Each Sylow 2-subgroup fixes exactly one of α,β, or γ, and therefore their intersection V is the subgroup of $\mathrm{Sym}(\{\alpha_1,\alpha_2,\alpha_3,\alpha_4\})$ fixing α, β, and γ.

LEMMA 4.8 *The fixed field of $G_f \cap V$ is $F[\alpha,\beta,\gamma]$. Hence $F[\alpha,\beta,\gamma]$ is Galois over F with Galois group $G_f/G_f \cap V$.*

PROOF. The above discussion shows that the subgroup of G_f of elements fixing $F[\alpha,\beta,\gamma]$ is $G_f \cap V$, and so $E^{G_f \cap V} = F[\alpha,\beta,\gamma]$ by the fundamental theorem of Galois theory. The remaining statements follow from the fundamental theorem using that V is normal. $\square$

$$
\begin{array}{c}
E \\
\bigg| \, {\scriptstyle G_f \cap V} \\
F[\alpha,\beta,\gamma] \\
\bigg| \, {\scriptstyle G_f/G_f \cap V} \\
F
\end{array}
$$

Let $M = F[\alpha, \beta, \gamma]$, and let $g(X) = (X - \alpha)(X - \beta)(X - \gamma) \in M[X]$ — it is called the **resolvent cubic** of f. Every permutation of the α_i (*a fortiori,* every element of G_f) permutes α, β, γ, and so fixes $g(X)$. Therefore (by the fundamental theorem) $g(X)$ has coefficients in F. More explicitly, the following is true.

LEMMA 4.9 *The resolvent cubic of* $f = X^4 + bX^3 + cX^2 + dX + e$ *is*

$$g = X^3 - cX^2 + (bd - 4e)X - b^2 e + 4ce - d^2.$$

The discriminants of f *and* g *are equal.*

SKETCH OF PROOF. Expand $f = (X - \alpha_1)(X - \alpha_2)(X - \alpha_3)(X - \alpha_4)$ to express b, c, d, e in terms of $\alpha_1, \alpha_2, \alpha_3, \alpha_4$. Expand $g = (X - \alpha)(X - \beta)(X - \gamma)$ to express the coefficients of g in terms of $\alpha_1, \alpha_2, \alpha_3, \alpha_4$, and substitute to express them in terms of b, c, d, e. □

Now let f be an irreducible separable quartic. Then $G = G_f$ is a transitive subgroup of S_4 whose order is divisible by 4. There are the following possibilities for G:

G	$(G \cap V : 1)$	$(G : V \cap G)$
S_4	4	6
A_4	4	3
V	4	1
D_4	4	2
C_4	2	2

$$E$$
$$\big|\, G \cap V$$
$$M$$
$$\big|\, G/G \cap V$$
$$F$$

The groups of type D_4 are the Sylow 2-subgroups discussed above, and the groups of type C_4 are those generated by cycles of length 4.

We can compute $(G : V \cap G)$ from the resolvent cubic g, because $G/V \cap G = \mathrm{Gal}(M/F)$ and M is the splitting field of g. Once we know $(G : V \cap G)$, we can deduce G except in the case that the index is 2. If $[M : F] = 2$, then $G \cap V = V$ or C_2. Only the first group acts transitively on the roots of f, and so (from 4.5) we see that in this case $G = D_4$ or C_4 according as f is irreducible or not in $M[X]$.

EXAMPLE 4.10 Consider $f(X) = X^4 - 4X + 2 \in \mathbb{Q}[X]$. It is irreducible by Eisenstein's criterion (1.16), and its resolvent cubic is $g(X) = X^3 - 8X - 16$, which is irreducible because it has no roots modulo 5. The discriminant of

$g(X)$ is -4864, which is not a square, and so the Galois group of $g(X)$ is S_3. From the table, we see that the Galois group of $f(X)$ is S_4.

EXAMPLE 4.11 Consider $f(X) = X^4 + 4X^2 + 2 \in \mathbb{Q}[X]$. It is irreducible by Eisenstein's criterion (1.16), and its resolvent cubic is $(X - 4)(X^2 - 8)$; thus $M = \mathbb{Q}[\sqrt{2}]$. From the table we see that G_f is of type D_4 or C_4, but f factors over M (even as a polynomial in X^2), and hence G_f is of type C_4.

EXAMPLE 4.12 Consider $f(X) = X^4 - 10X^2 + 4 \in \mathbb{Q}[X]$. It is irreducible in $\mathbb{Q}[X]$ because (by inspection) it is irreducible in $\mathbb{Z}[X]$. Its resolvent cubic is $(X + 10)(X + 4)(X - 4)$, and so G_f is of type V.

EXAMPLE 4.13 Consider $f(X) = X^4 - 2 \in \mathbb{Q}[X]$. It is irreducible by Eisenstein's criterion (1.16), and its resolvent cubic is $g(X) = X^3 + 8X$. Hence $M = \mathbb{Q}[i\sqrt{2}]$. One can check that f is irreducible over M, and G_f is of type D_4.

Alternatively, analyse the equation as in 3.23.

As we explained in 1.29, PARI knows how to factor polynomials with coefficients in $\mathbb{Q}[\alpha]$.

EXAMPLE 4.14 Consider $f(X) = X^4 - 2cX^3 - dX^2 + 2cdX - dc^2 \in \mathbb{Z}[X]$ with $a > 0$, $b > 0$, $c > 0$, $a > b$ and $d = a^2 - b^2$. Let $r = d/c^2$ and let w be the unique positive real number such that $r = w^3/(w^2 + 4)$. Let m be the number of roots of $f(X)$ in $\mathbb{Z}$ (counted with multiplicities). The Galois group of f is as follows:

- ◇ if $m = 0$ and w not rational, then G is S_4;

- ◇ if $m = 1$ and w not rational then G is S_3;

- ◇ if w is rational and $w^2 + 4$ is not a square then $G = D_4$;

- ◇ if w is rational and $w^2 + 4$ is a square then $G = V = C_2 \times C_2$.

This covers all possible cases. The hard part is to establish that $m = 2$ never happens.

Examples of polynomials with Galois group S_p over $\mathbb{Q}$

The next lemma gives a criterion for a subgroup of S_p to be the whole group.

LEMMA 4.15 *For p prime, the symmetric group S_p is generated by any transposition and any p-cycle.*

Proof. After renumbering, we may suppose that the transposition is $\tau = (12)$, and we may write the p-cycle σ so that 1 occurs in the first position, $\sigma = (1\, i_2 \cdots i_p)$. Now some power of σ will map 1 to 2 and will still be a p-cycle (here is where we use that p is prime). After replacing σ with the power, we have $\sigma = (1\, 2\, j_3 \ldots j_p)$, and after renumbering again, we have $\sigma = (1\, 2\, 3 \ldots p)$. Now

$$(i+1\ i+2) = \sigma^i (12)\sigma^{-i}$$

(see GT, 4.29) and so it lies in the subgroup generated by σ and τ. These transpositions generate S_p. $\qquad\square$

Proposition 4.16 *Let f be an irreducible polynomial of prime degree p in $\mathbb{Q}[X]$. If f splits in $\mathbb{C}$ and has exactly two nonreal roots, then $G_f = S_p$.*

Proof. Let E be the splitting field of f in $\mathbb{C}$, and let $\alpha \in E$ be a root of f. Because f is irreducible, $[\mathbb{Q}[\alpha]:\mathbb{Q}] = \deg f = p$, and so $p|[E:\mathbb{Q}] = (G_f:1)$. Therefore G_f contains an element of order p (Cauchy's theorem, GT, 4.13), but the only elements of order p in S_p are p-cycles (here we again use that p is prime).

Let σ be complex conjugation on $\mathbb{C}$. Then σ transposes the two nonreal roots of $f(X)$ and fixes the rest. Therefore $G_f \subset S_p$ and contains a transposition and a p-cycle, and so is the whole of S_p. $\qquad\square$

It remains to construct polynomials satisfying the conditions of the Proposition.

Example 4.17 *Let $p \geq 5$ be a prime number. Choose a positive even integer m and even integers*

$$n_1 < n_2 < \cdots < n_{p-2},$$

and let

$$g(X) = (X^2 + m)(X - n_1) \cdots (X - n_{p-2}).$$

The graph of g crosses the x-axis exactly at the points $n_1, \ldots, n_{p-2}$, and it does not have a local maximum or minimum at any of those points (because the n_i are simple roots). Thus $e = \min_{g'(x)=0} |g(x)| > 0$, and we can choose an odd positive integer n such that $\frac{2}{n} < e$.

Consider

$$f(X) = g(X) - \frac{2}{n}.$$

As $\frac{2}{n} < e$, the graph of f also crosses the x-axis at exactly $p-2$ points, and so f has exactly two nonreal roots. On the other hand, when we write

$$nf(X) = nX^p + a_1 X^{p-1} + \cdots + a_p,$$

the a_i are all even and a_p is not divisible by 2^2, and so Eisenstein's criterion implies that f is irreducible. Over $\mathbb{R}$, f has $p-2$ linear factors and one irreducible quadratic factor, and so it certainly splits over $\mathbb{C}$ (high school algebra). Therefore, the proposition applies to f.[1]

REMARK 4.18 The reader should not think that, in order to have Galois group S_p, a polynomial must have exactly two nonreal roots. For example, the polynomial $X^5 - 5X^3 + 4X - 1$ has Galois group S_5 but its roots are all real.

Finite fields

Let $\mathbb{F}_p = \mathbb{Z}/p\mathbb{Z}$, the field of p elements. As we noted in §1, every field E of characteristic p contains a copy of $\mathbb{F}_p$, namely, $\{m1_E \mid m \in \mathbb{Z}\}$. No harm results if we identify $\mathbb{F}_p$ with this subfield of E.

Let E be a field of degree n over $\mathbb{F}_p$. Then E has $q \stackrel{\text{def}}{=} p^n$ elements, and so $E^\times$ is a group of order $q-1$. Therefore the nonzero elements of E are roots of $X^{q-1} - 1$ (Lagrange's theorem, GT 1.27), and *all* elements of E including 0 are roots of $X^q - X$. Hence E is a splitting field for $X^q - X$, and so any two fields with q elements are isomorphic.

PROPOSITION 4.19 *Every extension of finite fields is simple.*

PROOF. Consider $E \supset F$. Then $E^\times$ is a finite subgroup of the multiplicative group of a field, and hence is cyclic (see Exercise 1-3). If ζ generates $E^\times$ as a multiplicative group, then certainly $E = F[\zeta]$.   □

Now let E be a splitting field of $f(X) = X^q - X$, $q = p^n$. As the derivative of f is the constant -1, which is relatively prime to f, we see that $f(X)$ has q distinct roots in E (2.21). Let S be the set of its roots. Then S is

---

[1] If m is taken sufficiently large, then $g(X) - 2$ will have exactly two nonreal roots, i.e., we can take $n = 1$, but the proof is longer (see Jacobson, *Lectures in Abstract Algebra*, 1964, Vol. III, p. 107, who credits the example to Brauer). The shorter argument in the text was suggested to me by Martin Ward.

obviously closed under multiplication and the formation of inverses, but it is also closed under subtraction: if $a^q = a$ and $b^q = b$, then

$$(a - b)^q = a^q - b^q = a - b.$$

Hence S is a field, and so $S = E$. In particular, E has q elements.

PROPOSITION 4.20 *For each power $q = p^n$ of p there exists a field $\mathbb{F}_q$ with q elements. Every such field is a splitting field for $X^q - X$ over $\mathbb{F}_p$, and so any two are isomorphic. Moreover, $\mathbb{F}_q$ is Galois over $\mathbb{F}_p$ with cyclic Galois group generated by the Frobenius automorphism $\sigma(a) = a^p$.*

PROOF. Only the final statement remains to be proved. The field $\mathbb{F}_q$ is Galois over $\mathbb{F}_p$ because it is the splitting field of a separable polynomial. We noted in 1.4 that $x \overset{\sigma}{\mapsto} x^p$ is an automorphism of $\mathbb{F}_q$. An element a of $\mathbb{F}_q$ is fixed by σ if and only if $a^p = a$, but $\mathbb{F}_p$ consists exactly of such elements, and so the fixed field of $\langle \sigma \rangle$ is $\mathbb{F}_p$. This proves that $\mathbb{F}_q$ is Galois over $\mathbb{F}_p$ and that $\langle \sigma \rangle = \mathrm{Gal}(\mathbb{F}_q/\mathbb{F}_p)$ (see 3.11). $\qquad\square$

COROLLARY 4.21 *Let E be a field with p^n elements. For each positive divisor m of n, E contains exactly one field with p^m elements.*

PROOF. We know that E is Galois over $\mathbb{F}_p$ and that $\mathrm{Gal}(E/\mathbb{F}_p)$ is the cyclic group of order n generated by σ. The group $\langle \sigma \rangle$ has one subgroup of order n/m for each m dividing n, namely, $\langle \sigma^m \rangle$, and so E has exactly one subfield of degree m over $\mathbb{F}_p$ for each m dividing n, namely, $E^{\langle \sigma^m \rangle}$. Because it has degree m over $\mathbb{F}_p$, $E^{\langle \sigma^m \rangle}$ has p^m elements. $\qquad\square$

COROLLARY 4.22 *Let $f \in \mathbb{F}_p[X]$ be a monic irreducible of degree d. If $d \mid n$, then f occurs exactly once as a factor of $X^{p^n} - X$. The degree of the splitting field of f is $\leq d$.*

PROOF. The factors of $X^{p^n} - X$ are distinct because it has no common factor with its derivative.[2] As $f(X)$ is irreducible of degree d, it has a root in a field of degree d over $\mathbb{F}_p$. But the splitting field of $X^{p^n} - X$ contains a copy of every field of degree d over $\mathbb{F}_p$ with $d \mid n$. Hence some root of $X^{p^n} - X$ is also a root of $f(X)$, and therefore $f(X) \mid X^{p^n} - X$. This proves the first statement. For the second, f divides $X^{p^d} - X$, and therefore it splits in its splitting field, which has degree d over $\mathbb{F}_p$. $\qquad\square$

---

[2] If $h(X) = f(X)^2 g(X)$, then $h'(X) = 2f'(X)f(X)g(X) + f(X)^2 g(X)$.

PROPOSITION 4.23 *Let $\mathbb{F}$ be an algebraic closure of $\mathbb{F}_p$. Then $\mathbb{F}$ contains exactly one field $\mathbb{F}_{p^n}$ with p^n elements for each integer $n \geq 1$, and $\mathbb{F}_{p^n}$ consists of the roots of $X^{p^n} - X$. Moreover,*

$$\mathbb{F}_{p^m} \subset \mathbb{F}_{p^n} \iff m \mid n,$$

in which case, $\mathbb{F}_{p^n}$ is Galois over $\mathbb{F}_{p^m}$ with Galois group generated by $x \mapsto x^{p^m}$.

PROOF. In fact, the set of roots of $X^{p^n} - X$ is a field (see above) with p^n elements, and it is the only such subfield. If $\mathbb{F}_{p^m} \subset \mathbb{F}_{p^n}$, say, $[\mathbb{F}_{p^n} : \mathbb{F}_{p^m}] = d$, then $p^n = (p^m)^d = p^{md}$, and so $m \mid n$; the converse follows from the first statement. If $m \mid n$, then $\mathbb{F}_{p^m}$ is the fixed field of the group generated by the automorphism $x \mapsto x^{p^m}$ of $\mathbb{F}_{p^n}$, and so the final assertion follows from Artin's theorem (3.11). □

The proposition shows that the partially ordered set of finite subfields of $\mathbb{F}$ is isomorphic to the set of integers $n \geq 1$ partially ordered by divisibility.

PROPOSITION 4.24 *The field $\mathbb{F}_p$ has an algebraic closure $\mathbb{F}$.*

PROOF. Choose a sequence of integers $1 = n_1 < n_2 < n_3 < \cdots$ such that $n_i \mid n_{i+1}$ for all i, and every integer n divides some n_i. For example, let $n_i = i!$. Define the fields $\mathbb{F}_{p^{n_i}}$ inductively as follows: $\mathbb{F}_{p^{n_1}} = \mathbb{F}_p$; $\mathbb{F}_{p^{n_i}}$ is the splitting field of $X^{p^{n_i}} - X$ over $\mathbb{F}_{p^{n_{i-1}}}$. Then, $\mathbb{F}_{p^{n_1}} \subset \mathbb{F}_{p^{n_2}} \subset \mathbb{F}_{p^{n_3}} \subset \cdots$, and we set $\mathbb{F} = \bigcup \mathbb{F}_{p^{n_i}}$. As a union of a chain of fields algebraic over $\mathbb{F}_p$, it is again a field algebraic over $\mathbb{F}_p$. Moreover, every polynomial in $\mathbb{F}_p[X]$ splits in $\mathbb{F}$, and so it is an algebraic closure of $\mathbb{F}$ (by 1.45). □

REMARK 4.25 Since the $\mathbb{F}_{p^n}$ are not subsets of a fixed set, forming the union requires explanation. One can appeal to the Axiom of Union in Zermelo-Fraenkel set theory for its existence, or, more naively, let S be the disjoint union of the $\mathbb{F}_{p^n}$. For $a, b \in S$, set $a \sim b$ if $a = b$ in one of the $\mathbb{F}_{p^n}$. Then $\sim$ is an equivalence relation, and we let $\mathbb{F} = S / \sim$.

Any two fields with q elements are isomorphic, but not necessarily *canonically* isomorphic. However, once we have chosen an algebraic closure $\mathbb{F}$ of $\mathbb{F}_p$, there is a *unique* subfield of $\mathbb{F}$ with q elements.

PARI factors polynomials modulo p very quickly. Recall that the syntax is `factormod(f(X),p)`. For example, to obtain a list of all monic polynomials of degree $1, 2$, or 4 over $\mathbb{F}_5$, ask PARI to factor $X^{625} - X$ modulo 5 (note that $625 = 5^4$).

NOTES In one of the few papers published during his short lifetime, entitled "Sur la theorie des nombres", which appeared in the Bulletin des Sciences Mathématiques in June 1830, Galois—at that time not even nineteen years old—defined finite fields of arbitrary prime power order and established their basic properties, e.g. the existence of a primitive element. So it is fully justified when finite fields are called **Galois fields** and customarily denoted by GF(q). (Letter, Notices A.M.S., Feb. 2003, p. 198, Péter P. Pálfry.)

Computing Galois groups over $\mathbb{Q}$

In this section, I describe a practical method for computing Galois groups over $\mathbb{Q}$ and similar fields. Recall that for a separable polynomial $f \in F[X]$, F_f denotes a splitting field for F, and $G_f = \mathrm{Gal}(F_f/F)$ denotes the Galois group of f. Moreover, G_f permutes the roots $\alpha_1, \ldots, \alpha_m$, $m = \deg f$, of f in F_f:

$$G \subset \mathrm{Sym}\{\alpha_1, \ldots, \alpha_m\}.$$

The first result generalizes Proposition 4.5.

PROPOSITION 4.26 *Let $f(X)$ be a separable polynomial in $F[X]$, and suppose that the orbits of G_f acting on the roots of f have $m_1, \ldots, m_r$ elements respectively. Then f factors as $f = f_1 \cdots f_r$ with f_i irreducible of degree m_i.*

PROOF. We may suppose that f is monic. Let $\alpha_1, \ldots, \alpha_m$, be the roots of $f(X)$ in F_f. The monic factors of $f(X)$ in $F_f[X]$ are in one-to-one correspondence with the subsets S of $\{\alpha_1, \ldots, \alpha_m\}$,

$$S \leftrightarrow f_S = \prod_{\alpha \in S} (X - \alpha).$$

Moreover, f_S is fixed under the action of G_f (and hence has coefficients in F) if and only if S is stable under G_f. Therefore the monic irreducible factors of f in $F[X]$ are the polynomials f_S corresponding to minimal subsets S of $\{\alpha_1, \ldots, \alpha_m\}$ stable under G_f, but these are precisely the orbits of G_f in $\{\alpha_1, \ldots, \alpha_m\}$.   □

REMARK 4.27 The proof shows the following more precise statement: let $\{\alpha_1, \ldots, \alpha_m\} = \bigcup O_i$ be the decomposition of $\{\alpha_1, \ldots, \alpha_m\}$ into a disjoint union of orbits for the group G_f; then $f = \prod f_i$, where $f_i = \prod_{\alpha_j \in O_i} (X - \alpha_j)$, is the decomposition of f into a product of monic irreducible polynomials in $F[X]$.

Now suppose that F is finite, with q elements say. Then G_f is a cyclic group generated by the Frobenius automorphism $\sigma: x \mapsto x^q$. When we regard σ as a permutation of the roots of f, then the orbits of σ correspond to the factors in its cycle decomposition (GT, 4.26). Hence, if the degrees of the distinct irreducible factors of f are $m_1, m_2, \ldots, m_r$, then σ has a cycle decomposition of type

$$m_1 + \cdots + m_r = \deg f.$$

THEOREM 4.28 (DEDEKIND) *Let $f(X) \in \mathbb{Z}[X]$ be a monic polynomial of degree m, and let p be a prime number such that f mod p has simple roots (equivalently, $D(f)$ is not divisible by p). Suppose that $\bar{f} = \prod_{i=1}^{r} f_i$ with f_i irreducible of degree m_i in $\mathbb{F}_p[X]$. Then G_f contains an element σ_f which, when viewed as a permutation of the roots of f, has a cycle decomposition $\sigma_1 \cdots \sigma_r$ with σ_i of length m_i.*

PROOF. Let $\alpha_1, \ldots, \alpha_m$ be the roots of f in some splitting field E_f of f, and let $A = \mathbb{Z}[\alpha_1, \ldots, \alpha_m]$. Clearly A is finitely generated as a $\mathbb{Z}$-module, and so p is not invertible in A. Therefore, it is contained in a maximal ideal P of A,[3] and $P \cap \mathbb{Z} = p\mathbb{Z}$. We shall show that G_f contains a unique element σ_P such that $\sigma_P(a) \equiv a^p$ mod P for all $a \in A$ (in particular, $\sigma_P(P) = P$).

Write $a \mapsto \bar{a}$ for the quotient map $A \to A/P$, and let $\bar{f} = f$ mod p. The quotient $A/P = \mathbb{F}_p[\bar{\alpha}_1, \ldots, \bar{\alpha}_m]$ is a splitting field $E_{\bar{f}}$ of $\bar{f}$. The group $G_{\bar{f}} \stackrel{\text{def}}{=} \mathrm{Gal}(E_{\bar{f}}/\mathbb{F}_p)$ is cyclic with generator $\bar{a} \mapsto \bar{a}^p$ (see 4.20). Let

$$D_P = \{\sigma \in G_f \mid \sigma(P) = P\}.$$

It is a subgroup of G_f. Each $\sigma \in D_P$ defines an automorphism $\bar{\sigma}$ of $E_{\bar{f}} \stackrel{\text{def}}{=} A/P$. The homomorphism $\phi: D_P \to G_{\bar{f}}$, $\sigma \mapsto \bar{\sigma}$, is injective because σ is determined by its action on the α_i, and hence by its action on the $\bar{\alpha}_i$. We now show that it is surjective.

Let $a \in A \smallsetminus P$. According to the Chinese remainder theorem (see 8.1 below), there exists a $b \in A$ such that $b \equiv a$ mod P and $b \equiv 0$ mod $\sigma^{-1}(P)$ for all $\sigma \in G_f \smallsetminus D_P$. Let $g(X) = \prod_{\sigma \in G_f}(X - \sigma(b))$. Then $g(X)$ lies in $\mathbb{Z}[X]$ and $\bar{g}(X) = X^s \prod_{\sigma \in D_P}(X - \bar{\sigma}(\bar{a}))$, where $s = |G \smallsetminus D_P|$, lies in $\mathbb{F}_p[X]$. The minimal polynomial of $\bar{a}$ over $\mathbb{F}_p$ divides $\bar{g}(X)$. On choosing a so that $E_{\bar{f}} = \mathbb{F}_p[\bar{a}]$, we find that D_P has order $[E_f : \mathbb{F}_p]$, and so $D_P \simeq G_{\bar{f}}$.

Let σ_P be the element of D_P such that $\bar{\sigma}_P = (\bar{a} \mapsto \bar{a}^p)$. Then σ_P is the unique element of G_f such that $\sigma_P(a) \equiv a^p$ mod P for all $a \in A$. Since

---

[3]Let P be the inverse image of any proper ideal of $A/(p)$ of highest dimension (as an $\mathbb{F}_p$-vector space).

$a \mapsto \bar{a}$ maps the roots of f bijectively onto the roots of $\bar{f}$, we see that D_P and G_f are isomorphic when viewed as permutation groups. Thus the cycle decomposition of σ_f is as described. $\qquad\square$

For an alternative proof of Dedekind's theorem, see Van der Waerden, *Modern Algebra*, I, §61 (or v5.00 of these notes).

ASIDE 4.29 Let E be a finite Galois extension of $\mathbb{Q}$ with Galois group G, and let $\mathcal{O}_E$ be the ring integers in E, i.e., the set of elements of E satisfying a monic polynomial in $\mathbb{Z}[X]$. Let P be a prime ideal of $\mathcal{O}_E$ such that $P \cap \mathbb{Z} = p\mathbb{Z}$. As in the above proof, there exists a unique element $\sigma_P \in G$ such that $\sigma_P P = P$ and $\sigma_P(a) \equiv a^p \bmod P$ for all $a \in \mathcal{O}_E$. This is called the **Frobenius automorphism** at P. If Q is a second prime ideal of $\mathcal{O}_E$ such that $Q \cap \mathbb{Z} = p\mathbb{Z}$, then $Q = \tau P$ for some $\tau \in G$, and $\sigma_Q = \tau \circ \sigma_P \circ \tau^{-1}$. The conjugacy class of σ_P is called the **Frobenius class** at p. When G is abelian, it consists of a single element.

EXAMPLE 4.30 Consider $X^5 - X - 1$. Modulo 2, this factors as

$$(X^2 + X + 1)(X^3 + X^2 + 1),$$

and modulo 3 it is irreducible. The theorem shows that G_f contains permutations $(ik)(lmn)$ and (12345), and so also $((ik)(lmn))^3 = (ik)$. Therefore $G_f = S_5$ by (4.15).

LEMMA 4.31 *A transitive subgroup of $H \subset S_n$ containing a transposition and an $(n-1)$-cycle is equal to S_n.*

PROOF. After renumbering, we may suppose that the $(n-1)$-cycle is the cycle $(123\ldots n-1)$. Because of the transitivity, the transposition can be transformed into (in), some $1 \leq i \leq n-1$. Conjugating (in) by $(123\ldots n-1)$ and its powers will transform it into $(1n)$, $(2n)$, ..., $(n-1\,n)$, and these elements obviously generate S_n. $\qquad\square$

EXAMPLE 4.32 Select separable monic polynomials of degree n, f_1, f_2, f_3 with coefficients in $\mathbb{Z}$ with the following factorizations:

  (a) f_1 is irreducible modulo 2;

  (b) $f_2 = $ (degree 1)(irreducible of degree $n-1$) mod 3;

  (c) $f_3 = $ (irreducible of degree 2)(product of 1 or 2 irreducible polynomials of odd degree) mod 5.

Take

$$f = -15 f_1 + 10 f_2 + 6 f_3.$$

Then

(i) G_f is transitive (it contains an n-cycle because $f \equiv f_1 \bmod 2$);

(ii) G_f contains a cycle of length $n-1$ (because $f \equiv f_2 \bmod 3$);

(iii) G_f contains a transposition (because $f \equiv f_3 \bmod 5$, and so it contains the product of a transposition with a commuting element of odd order; on raising this to an appropriate odd power, we are left with the transposition). Hence G_f is S_n.

The above results give the following strategy for computing the Galois group of an irreducible polynomial $f \in \mathbb{Q}[X]$. Factor f modulo a sequence of primes p not dividing $D(f)$ to determine the cycle types of the elements in G_f — a difficult theorem in number theory, the effective Chebotarev density theorem, says that if a cycle type occurs in G_f, then this will be seen by looking modulo a set of prime numbers of positive density, and will occur for a prime less than some bound. Now look up a table of transitive subgroups of S_n with order divisible by n and their cycle types. If this does not suffice to determine the group, then look at its action on the set of subsets of r roots for some r.

In Butler and McKay, *The transitive groups of degree up to eleven*, Comm. Algebra 11 (1983), 863–911, there is a list of all transitive subgroups of S_n, $n \le 11$, together with the cycle types of their elements and the orbit lengths of the subgroup acting on the r-sets of roots. With few exceptions, these invariants are sufficient to determine the subgroup up to isomorphism. See also, Soicher and McKay, *Computing Galois groups over the rationals*, J. Number Theory, 20 (1985) 273–281.

PARI can compute Galois groups for polynomials of degree ≤ 11 over $\mathbb{Q}$. The syntax is `polgalois(f)`, where f is an irreducible polynomial of degree ≤ 11, and the output is (n,s,k,name), where n is the order of the group, s is $+1$ or -1 according as the group is a subgroup of the alternating group or not, and "name" is the name of the group. For example, `polgalois(X^5-5*X^3+4*X-1)` (see 4.18) returns the symmetric group S_5, which has order 120, `polgalois(X^11-5*X^3+4*X-1)` returns the symmetric group S_{11}, which has order 39916800, and `polgalois(X^12-5*X^3...)` returns an apology. The reader should use PARI to check the examples 4.10–4.13.

ASIDE 4.33 For a monic polynomial f of degree n with bounded integers as coefficients, it is expected that the Galois group of f equals S_n with probability 1 as $n \to \infty$. See Bary-Soroker, Kozma, and Gady, Duke Math. J. 169 (2020), 579–598, for precise statements.

ASIDE 4.34 Of the $(2H+1)^n$ monic polynomials $f(X) = X^n + a_1 X^{n-1} + \cdots + a_n \in \mathbb{Z}[X]$ with $\max\{|a_1|, \ldots, |a_n|\} = H$, how many have Galois group $\neq S_n$? There are clearly $\gg H^{n-1}$ such polynomials, as may be seen by setting $a_n = 0$. It was conjectured by Van der Waerden in 1936, and proved by Bhargava in 2021, that $O(H^{n-1})$ is in fact be the correct upper bound for the count of such polynomials.

Appendix: Computing discriminants using resultants

Let $f, g \in F[X]$, and suppose that

$$f(X) = a \prod_1^n (X - \alpha_i), \quad g(X) = b \prod_1^m (X - \beta_j), \quad ab \neq 0,$$

in some splitting field for fg. The **resultant** of f and g is defined by

$$\mathrm{Res}(f, g) = a^m b^n \prod_{i,j} (\alpha_i - \beta_j).$$

PROPOSITION 4.35 *Let $f, g \in F[X]$ as above. Then,*
  (a) $\mathrm{Res}(f, g) = (-1)^{mn} \mathrm{Res}(g, f)$;
  (b) $\mathrm{Res}(f, g) = a^m \prod_{i=1}^n g(\alpha_i)$;
  (c) *If $g \equiv g_1 \bmod f$ in $F[X]$ with $\deg(g_1) = m_1$, then*

$$\mathrm{Res}(f, g) = a^{m - m_1} \mathrm{Res}(f, g_1).$$

PROOF. Statements (a) and (b) are obvious. If $g \equiv g_1 \bmod f$, then

$$\prod_{i=1}^n g(\alpha_i) = \prod_{i=1}^n g_1(\alpha_i),$$

and so (c) follows from (b) .   □

These formulas make it possible to compute resultants by applying the division algorithm to reduce the degree of g, then switching the two polynomials, and continuing until one polynomial has degree ≤ 1.

PROPOSITION 4.36 *Let $f \in F[X]$ be a monic polynomial of degee n, and let f' be its derivative. Then*

$$D(f) = (-1)^{\frac{n(n-1)}{2}} \mathrm{Res}(f, f') = (-1)^{\frac{n(n-1)}{2}} \mathrm{Res}(f', f).$$

PROOF. If $f(X) = \prod_{i=1}^{n}(X - \alpha_i)$, then

$$D(f) \overset{\text{def}}{=} \prod_{1 \le i < j \le n}(\alpha_i - \alpha_j)^2$$

$$= (-1)^{\frac{n(n-1)}{2}} \prod_{i \ne j}(\alpha_i - \alpha_j)$$

$$= (-1)^{\frac{n(n-1)}{2}} \prod_{i=1}^{n}\prod_{j \ne i}(\alpha_i - \alpha_j).$$

On the other hand,

$$f'(X) = \sum_{i=1}^{n}\prod_{j \ne i}(X - \alpha_j),$$

and so $f'(\alpha_i) = \prod_{j \ne i}(\alpha_i - \alpha_j)$ for $i = 1,\ldots,n$. Now the statement follows from 4.35(b). $\qquad\square$

EXAMPLE 4.37 Let $f = X^3 + bX + c$. Then

$$D(f) = -\operatorname{Res}(3X^2 + b, X^3 + bX + c)$$

$$= -3^2 \operatorname{Res}(3X^2 + b, \frac{2b}{3}X + c)$$

because

$$X^3 + bX + c = \frac{X}{3}(3X^2 + b) + \frac{2b}{3}X + c.$$

Thus

$$D(f) = -3^2 \operatorname{Res}\left(\frac{2b}{3}(X + \frac{3c}{2b}), 3X^2 + b\right) \qquad \text{by 4.35(a)}$$

$$= -3^2 \cdot \left(\frac{2b}{3}\right)^2 \left(3\left(\frac{3c}{2b}\right)^2 + b\right) \qquad \text{by 4.35(b)}$$

$$= -4b^3 - 27c^2.$$

EXAMPLE 4.38 Let $f = X^5 + X + 1$. Then

$$D(f) = \operatorname{Res}(5X^4 + 1, X^5 + X + 1)$$

$$= 5^4 \operatorname{Res}(5X^4 + 1, \frac{4}{5}X + 1)$$

because

$$X^5 + X + 1 = \frac{X}{5}(5X^4 + 1) + \frac{4}{5}X + 1.$$

Thus

$$D(f) = 5^4 \operatorname{Res}\left(\frac{4}{5}(X + \frac{5}{4}), 5X^4 + 1\right) \qquad \text{by 4.35(a)}$$

$$= 5^4 \left(\frac{4}{5}\right)^4 \left(5\left(\frac{5}{4}\right)^4 + 1\right) \qquad \text{by 4.35(b)}$$

$$= 3381.$$

EXAMPLE 4.39 Let $f = X^n + aX + b$. Then

$$D(f) = (-1)^{\frac{n(n-1)}{2}} \operatorname{Res}(nX^{n-1} + a, X^n + aX + b)$$

$$= (-1)^{\frac{n(n-1)}{2}} n^{n-1} \operatorname{Res}(nX^{n-1} + a, a\frac{n-1}{n}X + b)$$

because

$$X^n + aX + b = \frac{X}{n}(nX^{n-1} + a) + a\frac{n-1}{n}X + b.$$

Thus

$$D(f) = (-1)^{\frac{n(n-1)}{2}}(-n)^{n-1} \operatorname{Res}\left(a\frac{n-1}{n}(X + \frac{nb}{(n-1)a}), nX^{n-1} + a\right) \quad (4.35\text{a})$$

$$= (-1)^{\frac{n(n-1)}{2}}(-n)^{n-1}\left(a\frac{n-1}{n}\right)^{n-1}\left(n\left(\frac{-nb}{(n-1)a}\right)^{n-1} + a\right) \quad (4.35\text{b})$$

$$= (-1)^{\frac{n(n-1)}{2}}(-a(n-1))^{n-1}\left(n\left(\frac{-nb}{(n-1)a}\right)^{n-1} + a\right)$$

$$= (-1)^{\frac{n(n-1)}{2}}(n^n b^{n-1} + (-1)^{n-1}(n-1)^{n-1}a^n).$$

NOTES The appendix is based on a letter of René Schoof.

Exercises

4-1 Find the splitting field of $X^m - 1 \in \mathbb{F}_p[X]$.

4-2 Find the Galois group of $X^4 - 2X^3 - 8X - 3$ over $\mathbb{Q}$.

4-3 Find the degree of the splitting field of $X^8 - 2$ over $\mathbb{Q}$.

4-4 Give an example of a field extension E/F of degree 4 such that there does not exist a field M with $F \subset M \subset E$, $[M:F] = 2$.

4-5 List all irreducible polynomials of degree 3 over $\mathbb{F}_7$ in 10 seconds or less (there are 112).

4-6 "It is a thought-provoking question that few graduate students would know how to approach the question of determining the Galois group of, say,

$$X^6 + 2X^5 + 3X^4 + 4X^3 + 5X^2 + 6X + 7."$$

[over $\mathbb{Q}$].

(a) Can you find it?

(b) Can you find it without using the "polgalois" command in PARI?

4-7 Let $f(X) = X^5 + aX + b$, $a, b \in \mathbb{Q}$. Show that $G_f \approx D_5$ (dihedral group) if and only if

(a) $f(X)$ is irreducible in $\mathbb{Q}[X]$, and

(b) the discriminant $D(f) = 4^4 a^5 + 5^5 b^4$ of $f(X)$ is a square, and

(c) the equation $f(X) = 0$ is solvable by radicals.

4-8 Show that a polynomial f of degree $n = \prod_{i=1}^{k} p_i^{r_i}$ (the p_i are distinct primes) is irreducible over $\mathbb{F}_p$ if and only if (a) $\gcd(f(X), X^{p^{n/p_i}} - X) = 1$ for all $1 \leq i \leq k$ and (b) f divides $X^{p^n} - X$ (Rabin irreducibility test[4]).

4-9 Let $f(X)$ be an irreducible polynomial in $\mathbb{Q}[X]$ with both real and nonreal roots. Show that its Galois group is nonabelian. Can the condition that f is irreducible be dropped?

4-10 Let F be a Galois extension of $\mathbb{Q}$, and let α be an element of F such that $\alpha F^{\times 2}$ is not fixed by the action of $\mathrm{Gal}(F/\mathbb{Q})$ on $F^{\times}/F^{\times 2}$. Let $\alpha = \alpha_1, \ldots, \alpha_n$ be the orbit of α under $\mathrm{Gal}(F/\mathbb{Q})$. Show:

(a) $F[\sqrt{\alpha_1}, \ldots, \sqrt{\alpha_n}]/F$ is Galois with commutative Galois group contained in $(\mathbb{Z}/2\mathbb{Z})^n$.

(b) $F[\sqrt{\alpha_1}, \ldots, \sqrt{\alpha_n}]/\mathbb{Q}$ is Galois with noncommutative Galois group contained in $(\mathbb{Z}/2\mathbb{Z})^n \rtimes \mathrm{Gal}(F/\mathbb{Q})$.

---

[4]Rabin, *Probabilistic algorithms in finite fields.* SIAM J. Comput. 9 (1980), no. 2, 273–280.

CHAPTER 5

Applications of Galois Theory

In this chapter, we apply the fundamental theorem of Galois theory to obtain other results about polynomials and extensions of fields.

Primitive element theorem.

Recall that a finite extension of fields E/F is simple if $E = F[\alpha]$ for some element α of E. Such an α is called a ***primitive element*** of E. We'll show that (at least) all separable extensions have primitive elements.

Consider for example $\mathbb{Q}[\sqrt{2}, \sqrt{3}]/\mathbb{Q}$. We know (see Exercise 3-3) that its Galois group over $\mathbb{Q}$ is a 4-group $\langle \sigma, \tau \rangle$, where

$$\begin{cases} \sigma\sqrt{2} = -\sqrt{2} \\ \sigma\sqrt{3} = \sqrt{3} \end{cases}, \quad \begin{cases} \tau\sqrt{2} = \sqrt{2} \\ \tau\sqrt{3} = -\sqrt{3} \end{cases}.$$

Note that

$$\begin{aligned} \sigma(\sqrt{2}+\sqrt{3}) &= -\sqrt{2}+\sqrt{3}, \\ \tau(\sqrt{2}+\sqrt{3}) &= \sqrt{2}-\sqrt{3}, \\ (\sigma\tau)(\sqrt{2}+\sqrt{3}) &= -\sqrt{2}-\sqrt{3}. \end{aligned}$$

These all differ from $\sqrt{2}+\sqrt{3}$, and so only the identity element of the Galois group fixes all elements of $\mathbb{Q}[\sqrt{2}+\sqrt{3}]$. According to the fundamental theorem, this implies that $\sqrt{2}+\sqrt{3}$ is a primitive element:

$$\mathbb{Q}[\sqrt{2}, \sqrt{3}] = \mathbb{Q}[\sqrt{2}+\sqrt{3}].$$

It is clear that this argument should work much more generally.

THEOREM 5.1 *Let $E = F[\alpha_1, \ldots, \alpha_r]$ be a finite extension of F, and assume that $\alpha_2, \ldots, \alpha_r$ are separable over F (but not necessarily α_1). Then there exists a $\gamma \in E$ such that $E = F[\gamma]$.*

PROOF. For finite fields, we proved this in 4.19. Hence we may assume F to be infinite. It suffices to prove the statement for $r = 2$, for then

$$F[\alpha_1, \alpha_2, \ldots, \alpha_r] = F[\alpha_1', \alpha_3, \ldots, \alpha_r] = F[\alpha_1'', \alpha_4, \ldots, \alpha_r] = \cdots.$$

Thus let $E = F[\alpha, \beta]$ with β separable over F. Let f and g be the minimal polynomials of α and β over F, and let L be a splitting field for fg containing E. Let $\alpha_1 = \alpha, \ldots, \alpha_s$ be the roots of f in L, and let $\beta_1 = \beta, \beta_2, \ldots, \beta_t$ be the roots of g. For $j \neq 1$, $\beta_j \neq \beta$, and so the the equation

$$\alpha_i + X\beta_j = \alpha + X\beta,$$

has exactly one solution, namely, $X = \frac{\alpha_i - \alpha}{\beta - \beta_j}$. If we choose a $c \in F$ different from any of these solutions (using that F is infinite), then

$$\alpha_i + c\beta_j \neq \alpha + c\beta \text{ unless } i = 1 = j.$$

Let $\gamma = \alpha + c\beta$. I claim that

$$F[\alpha, \beta] = F[\gamma].$$

The polynomials $g(X)$ and $f(\gamma - cX)$ have coefficients in $F[\gamma]$, and have β as a root:

$$g(\beta) = 0, \quad f(\gamma - c\beta) = f(\alpha) = 0.$$

In fact, β is their only common root, because we chose c so that $\gamma - c\beta_j \neq \alpha_i$ unless $i = 1 = j$. Therefore

$$\gcd(g(X), f(\gamma - cX)) = X - \beta.$$

Here we computed the gcd in $L[X]$, but this is equal to the gcd computed in $F[\gamma][X]$ (Proposition 2.17). Hence $\beta \in F[\gamma]$, and this implies that $\alpha = \gamma - c\beta$ also lies in $F[\gamma]$. This proves the claim. $\square$

REMARK 5.2 When F is infinite, the proof shows that γ can be chosen to be of the form

$$\gamma = \alpha_1 + c_2\alpha_2 + \cdots + c_r\alpha_r, \quad c_i \in F.$$

If $F[\alpha_1, \ldots, \alpha_r]$ is Galois over F, then an element of this form will be a primitive element provided it is moved by every nontrivial element of the Galois group. This remark makes it very easy to write down primitive elements.

Our hypotheses are minimal: if *two* of the α are not separable, then the extension need not be simple. Before giving an example to illustrate this, we need another result.

PROPOSITION 5.3 *Let $E = F[\gamma]$ be a simple algebraic extension of F. Then there are only finitely many intermediate fields M,*

$$F \subset M \subset E.$$

PROOF. Let M be such a field, and let $g(X)$ be the minimal polynomial of γ over M. Let M' be the subfield of E generated over F by the coefficients of $g(X)$. Clearly $M' \subset M$, but (equally clearly) $g(X)$ is the minimal polynomial of γ over M'. Hence

$$[E:M'] = \deg(g) = [E:M],$$

and so $M = M'$; we have shown that M is generated by the coefficients of $g(X)$.

Let $f(X)$ be the minimal polynomial of γ over F. Then $g(X)$ divides $f(X)$ in $M[X]$, and hence also in $E[X]$. Therefore, there are only finitely many possible g, and consequently only finitely many possible M. $\square$

Note that the proof in fact gives a description of all the intermediate fields: each is generated over F by the coefficients of a factor $g(X)$ of $f(X)$ in $E[X]$. The coefficients of such a $g(X)$ are partially symmetric polynomials in the roots of $f(X)$ (that is, fixed by some, but not necessarily all, of the permutations of the roots).

REMARK 5.4 The proposition has a converse: If E is a finite extension of F and there are only finitely many intermediate fields M, $F \subset M \subset E$, then E is a simple extension of F.

In proving this, we may suppose that F is infinite, and use that no finite-dimensional F-vector space is a finite union of proper subspaces.[1] Thus there is an element γ in E not contained in any proper subfield, and so $E = F[\gamma]$. This gives another proof of Theorem 5.1 in the case that E is separable over F, because Galois theory shows that there are only finitely many intermediate fields in this case (even the Galois closure of E over F has only finitely many intermediate fields).

---

[1] Let $U_1, \ldots, U_m$ be proper subspaces of V, and let $f_1, \ldots, f_m$ be nonzero linear forms on V such that f_i is zero on U_i. If $V = \bigcup U_i$, then $f \stackrel{\text{def}}{=} f_1 \ldots f_m$ is zero on V, which implies that it is zero (5.19) — contradiction.

EXAMPLE 5.5 The simplest nonsimple algebraic extension is $k(X,Y) \supset k(X^p, Y^p)$, where k is an algebraically closed field of characteristic p. Let $F = k(X^p, Y^p)$. For all $c \in k$, we have

$$k(X,Y) = F[X,Y] \supset F[X+cY] \supset F$$

with the degree of each extension equal to p. If

$$F[X+cY] = F[X+c'Y], \quad c \neq c',$$

then $F[X+cY]$ would contain both X and Y, which is impossible because $[k(X,Y):F] = p^2$. Hence there are infinitely many distinct intermediate fields.[2]

Alternatively, note that the degree of $k(X,Y)$ over $k(X^p, Y^p)$ is p^2, but if $\alpha \in k(X,Y)$, then $\alpha^p \in k(X^p, Y^p)$, and so α generates a field of degree at most p over $k(X^p, Y^p)$.

Fundamental Theorem of Algebra

We finally prove the fundamental theorem of algebra.[3]

THEOREM 5.6 *The field $\mathbb{C}$ of complex numbers is algebraically closed.*

PROOF. We'll need to use the following two facts about $\mathbb{R}$:

◇　positive real numbers have square roots;

◇　every polynomial of odd degree with real coefficients has a real root.

Both are immediate consequences of the Intermediate Value Theorem, which says that a continuous function on a closed interval takes every value between its maximum and minimal values (inclusive).

We define $\mathbb{C}$ to be the splitting field of $X^2 + 1$ over $\mathbb{R}$, and we let i denote a root of $X^2 + 1$ in $\mathbb{C}$. Thus $\mathbb{C} = \mathbb{R}[i]$. We have to show (see 1.45) that every $f(X) \in \mathbb{R}[X]$ splits in $\mathbb{C}$. We may suppose that f is monic, irreducible, and $\neq X^2 + 1$.

---

[2]Zariski showed that there is even an intermediate field M that is not isomorphic to $F(X,Y)$, and Piotr Blass showed, using the methods of algebraic geometry, that there is an infinite sequence of intermediate fields, no two of which are isomorphic.

[3]This is not strictly a theorem in algebra: it is a statement about $\mathbb{R}$ whose construction is part of analysis (or maybe topology). In fact, I prefer the proof based on Liouville's theorem in complex analysis to the more algebraic proof given in the text: if $f(z)$ is a polynomial without a root in $\mathbb{C}$, then $f(z)^{-1}$ is bounded and holomorphic on the whole complex plane, and hence (by Liouville) constant.

We first show that every element α of $\mathbb{C}$ has a square root in $\mathbb{C}$. Write $\alpha = a + bi$, with $a, b \in \mathbb{R}$, and let c, d be real numbers such that

$$c^2 = \frac{a + \sqrt{a^2 + b^2}}{2}, \quad d^2 = \frac{-a + \sqrt{a^2 + b^2}}{2}.$$

Then $c^2 - d^2 = a$ and $(2cd)^2 = b^2$. If we choose the signs of c and d so that cd has the same sign as b, then $(c + di)^2 = \alpha$ and so $c + di$ is a square root of α.

Let $f(X) \in \mathbb{R}[X]$, and let E be a splitting field for $f(X)(X^2 + 1)$. Then E contains $\mathbb{C}$, and we have to show that it equals $\mathbb{C}$. Since $\mathbb{R}$ has characteristic zero, the polynomial is separable, and so E is Galois over $\mathbb{R}$ (see 3.10). Let G be its Galois group, and let H be a Sylow 2-subgroup of G.

Let $M = E^H$. Then M has of degree $(G:H)$ over $\mathbb{R}$, which is odd, and so the minimal polynomial over $\mathbb{R}$ of any $\alpha \in M$ has odd degree (by the multiplicativity of degrees, 1.20), and so has a real root. As it is irreducible, it has degree 1. Hence $\alpha \in \mathbb{R}$, and so $M = \mathbb{R}$ and $G = H$.

We deduce that $\mathrm{Gal}(E/\mathbb{C})$ is a 2-group. If it is $\neq 1$, then it has a subgroup N of index 2 (GT, 4.17). The field E^N has degree 2 over $\mathbb{C}$, and so it is generated by the square root of an element of $\mathbb{C}$ (see 3.25), but all square roots of elements of $\mathbb{C}$ lie in $\mathbb{C}$. Hence $E^N = \mathbb{C}$, which is a contradiction. Thus $\mathrm{Gal}(E/\mathbb{C}) = 1$ and $E = \mathbb{C}$. $\qquad\square$

COROLLARY 5.7 (a) *The field $\mathbb{C}$ is the algebraic closure of $\mathbb{R}$.*

(b) *The set of all algebraic numbers is an algebraic closure of $\mathbb{Q}$.*

PROOF. Part (a) is obvious from the definition of "algebraic closure" (1.44), and (b) follows from Corollary 1.47. $\qquad\square$

NOTES The Fundamental Theorem was quite difficult to prove. Gauss gave a proof in his doctoral dissertation in 1798 in which he used some geometric arguments which he did not justify. He gave the first rigorous proof in 1816. The elegant argument given here is a simplification by Emil Artin of earlier proofs (see Emil Artin, *Algebraische Konstruktion reeller Körper*, Hamb. Abh., Bd. 5 (1926), 85-90; translation available in Emil Artin, *Exposition by Emil Artin*. AMS; LMS 2007).

Cyclotomic extensions

A ***primitive*** nth root of 1 in F is an element of order n in $F^\times$. Such an element can exist only if the characteristic of F does not divide n (so either it is 0 or p not dividing n). We refer the reader to GT 3.5 for the group $(\mathbb{Z}/n\mathbb{Z})^\times$.

PROPOSITION 5.8 *Let F be a field of characteristic not dividing n, and let E be the splitting field of $X^n - 1$.*

(a) *There exists a primitive nth root of 1 in E.*

(b) *If ζ is a primitive nth root of 1 in E, then $E = F[\zeta]$.*

(c) *The field E is Galois over F; for each $\sigma \in \mathrm{Gal}(E/F)$, there is an $i \in (\mathbb{Z}/n\mathbb{Z})^{\times}$ such that $\sigma\zeta = \zeta^i$ for all ζ with $\zeta^n = 1$; the map $\sigma \mapsto [i]$ is an injective homomorphism*

$$\mathrm{Gal}(E/F) \to (\mathbb{Z}/n\mathbb{Z})^{\times}.$$

PROOF. (a) The roots of $X^n - 1$ are distinct, because its derivative nX^{n-1} has only zero as a root (here we use the condition on the characteristic), and so E contains n distinct nth roots of 1. The nth roots of 1 form a finite subgroup of $E^{\times}$, and so (see Exercise 3) they form a cyclic group. Every generator has order n, and hence is a primitive nth root of 1.

(b) The roots of $X^n - 1$ are the powers of ζ, and $F[\zeta]$ contains them all.

(c) The extension E/F is Galois because E is the splitting field of a separable polynomial. If ζ_0 is one primitive nth root of 1, then the remaining primitive nth roots of 1 are the elements ζ_0^i with i relatively prime to n. Since, for any automorphism σ of E, $\sigma\zeta_0$ is again a primitive nth root of 1, it equals ζ_0^i for some i relatively prime to n, and the map $\sigma \mapsto i \mod n$ is injective because ζ_0 generates E over F. It is obviously a homomorphism. Moreover, for any other nth root of 1, say, $\zeta = \zeta_0^m$, we have

$$\sigma\zeta = (\sigma\zeta_0)^m = \zeta_0^{im} = \zeta^i,$$

and so the homomorphism does not depend on the choice of ζ_0.   □

The map $\sigma \mapsto [i]: \mathrm{Gal}(F[\zeta]/F) \to (\mathbb{Z}/n\mathbb{Z})^{\times}$ need not be surjective. For example, if $F = \mathbb{C}$, then its image is $\{1\}$, and if $F = \mathbb{R}$, it is either $\{[1]\}$ or $\{[-1], [1]\}$. On the other hand, when $n = p$ is prime, we showed in Lemma 1.42 that $[\mathbb{Q}[\zeta]:\mathbb{Q}] = p - 1$, and so the map is surjective. We now prove that the map is surjective for all n when $F = \mathbb{Q}$.

The polynomial $X^n - 1$ has some obvious factors in $\mathbb{Q}[X]$, namely, the polynomials $X^d - 1$ for any $d \mid n$. When we remove all factors of $X^n - 1$ of this form with $d < n$, the polynomial we are left with is called the nth **cyclotomic polynomial** Φ_n. Thus

$$\Phi_n = \prod (X - \zeta) \qquad \text{(product over the primitive nth roots of 1)}.$$

It has degree $\varphi(n)$, the order of $(\mathbb{Z}/n\mathbb{Z})^{\times}$. Since every nth root of 1 is a primitive dth root of 1 for exactly one positive divisor d of n, we see that

$$X^n - 1 = \prod_{d\,|\,n} \Phi_d(X).$$

For example, $\Phi_1(X) = X - 1$, $\Phi_2(X) = X + 1$, $\Phi_3(X) = X^2 + X + 1$, and

$$\Phi_6(X) = \frac{X^6 - 1}{(X - 1)(X + 1)(X^2 + X + 1)} = X^2 - X + 1.$$

This gives an easy inductive method of computing the cyclotomic polynomials. Alternatively type `polcyclo(n,X)` in PARI.

Because $X^n - 1$ has coefficients in $\mathbb{Z}$ and is monic, every monic factor of it in $\mathbb{Q}[X]$ has coefficients in $\mathbb{Z}$ (see 1.14). In particular, the cyclotomic polynomials lie in $\mathbb{Z}[X]$.

LEMMA 5.9 *Let F be a field of characteristic not dividing n, and let ζ be a primitive nth root of 1 in some extension of F. The following are equivalent:*

(a) *the nth cyclotomic polynomial Φ_n is irreducible;*

(b) *the degree $[F[\zeta]\colon F] = \varphi(n)$;*

(c) *the homomorphism*

$$\mathrm{Gal}(F[\zeta]/F) \to (\mathbb{Z}/n\mathbb{Z})^{\times}$$

  is an isomorphism.

PROOF. Because ζ is a root of Φ_n, the minimal polynomial of ζ divides Φ_n. It equals it if and only if $[F[\zeta]\colon F] = \varphi(n)$, which is true if and only if the injection $\mathrm{Gal}(F[\zeta]/F) \hookrightarrow (\mathbb{Z}/n\mathbb{Z})^{\times}$ is onto. $\square$

THEOREM 5.10 *The nth cyclotomic polynomial Φ_n is irreducible in $\mathbb{Q}[X]$.*

PROOF. Let $f(X)$ be a monic irreducible factor of Φ_n in $\mathbb{Q}[X]$. Its roots will be primitive nth roots of 1, and we have to show they include *all* primitive nth roots of 1. For this it suffices to show that

$$\zeta \text{ a root of } f(X) \implies \zeta^i \text{ a root of } f(X) \text{ for all } i \text{ such that } \gcd(i,n) = 1.$$

Such an i is a product of primes not dividing n, and so it suffices to show that

$$\zeta \text{ a root of } f(X) \implies \zeta^p \text{ a root of } f(X) \text{ for all primes } p \text{ not dividing } n.$$

Write
$$\Phi_n(X) = f(X)g(X).$$

Proposition 1.14 shows that $f(X)$ and $g(X)$ lie in $\mathbb{Z}[X]$. Suppose that ζ is a root of f but that, for some prime p not dividing n, ζ^p is not a root of f. Then ζ^p is a root of $g(X)$, $g(\zeta^p) = 0$, and so ζ is a root of $g(X^p)$. As $f(X)$ and $g(X^p)$ have a common root, they have a nontrivial common factor in $\mathbb{Q}[X]$ (2.17), which automatically lies in $\mathbb{Z}[X]$ (1.14).

Write $h(X) \mapsto \bar{h}(X)$ for the quotient map $\mathbb{Z}[X] \to \mathbb{F}_p[X]$, and note that, because $f(X)$ and $g(X^p)$ have a common factor of degree ≥ 1 in $\mathbb{Z}[X]$, so also do $\bar{f}(X)$ and $\bar{g}(X^p)$ in $\mathbb{F}_p[X]$. The mod p binomial theorem shows that

$$\bar{g}(X)^p = \bar{g}(X^p)$$

(recall that $a^p = a$ for all $a \in \mathbb{F}_p$), and so $\bar{f}(X)$ and $\bar{g}(X)$ have a common factor of degree ≥ 1 in $\mathbb{F}_p[X]$. Hence $X^n - 1$, when regarded as an element of $\mathbb{F}_p[X]$, has multiple roots, but, as $p \nmid n$, it is separable — contradiction.

ALTERNATIVE PROOF. We have to show that the homomorphism

$$\phi \colon \mathrm{Gal}(\mathbb{Q}[\zeta]/\mathbb{Q}) \to (\mathbb{Z}/n\mathbb{Z})^\times$$

is surjective. Let $\sigma_p \in \mathrm{Gal}(\mathbb{Q}[\zeta]/\mathbb{Q})$ be the Frobenius class at a prime p not dividing n (see 4.29). Then $\sigma_p(\zeta) = \zeta^p$ because this is the only nth root of 1 congruent to ζ^p modulo a prime ideal lying over p, and so $\phi(\sigma_p) = [p]$. As $(\mathbb{Z}/n\mathbb{Z})^\times$ is generated by the classes of the prime numbers not dividing n, this shows that ϕ is surjective. $\qquad\square$

ASIDE 5.11 The proof of 5.10 is very old — in essence it goes back to Dedekind in 1857 — but its general scheme has recently become popular: take a statement in characteristic zero, reduce modulo p (where the statement may no longer be true), and exploit the existence of the Frobenius automorphism $a \mapsto a^p$ to obtain a proof of the original statement. For example, commutative algebraists use this method to prove results about commutative rings, and there are theorems about complex manifolds that were first proved by reducing things to characteristic p.

There are some beautiful relations between what happens in characteristic 0 and in characteristic p. For example, let $f(X_1, ..., X_n) \in \mathbb{Z}[X_1, ..., X_n]$. We can

(a) look at the solutions of $f = 0$ in $\mathbb{C}$, and so get a topological space;

(b) reduce mod p, and look at the solutions of $\bar{f} = 0$ in $\mathbb{F}_{p^n}$.

The Weil conjectures (Weil 1949; proved by Deligne, Grothendieck, ...) assert that the Betti numbers of the space in (a) control the cardinalities of the sets in (b).

THEOREM 5.12 *The regular n-gon is constructible if and only if n is of the form $2^k p_1 \cdots p_s$, where the p_i are distinct Fermat primes.*

PROOF. The regular n-gon is constructible if and only if $\cos\frac{2\pi}{n}$ (equivalently, $\zeta = e^{2\pi i/n}$) is constructible. We know that $\mathbb{Q}[\zeta]$ is Galois over $\mathbb{Q}$, and so (according to 1.38 and 3.24) ζ is constructible if and only if $[\mathbb{Q}[\zeta]:\mathbb{Q}]$ is a power of 2. When we write $n = \prod p^{n(p)}$,

$$\varphi(n) = \prod_{p\mid n}(p-1)p^{n(p)-1},$$

(GT, 3.5), and this is a power of 2 if and only if n has the required form. $\quad\square$

REMARK 5.13 As mentioned earlier, the Fermat primes are those of the form $2^{2^r}+1$. Because the Fermat primes are not known, the problem of listing the n for which the regular n-gon is constructible has not yet solved.

NOTES The final section of Gauss's, *Disquisitiones Arithmeticae* (1801) is titled "Equations defining sections of a Circle". In it Gauss proves that the nth roots of 1 form a cyclic group, that $X^n - 1$ is solvable (this was before the theory of abelian groups had been developed, and before Galois), and that the regular n-gon is constructible when n is as in the Theorem. He also claimed to have proved the converse statement. This leads some people to credit him with the above proof of the irreducibility of Φ_n, but in the absence of further evidence, I'm sticking with Dedekind. For a recent article discussing this, see Anderson, Chahal, and Top, *The last chapter of the Disquisitiones of Gauss*. Hardy–Ramanujan J. 44 (2021), 152–159, arXiv:2110.01355.

Dedekind's theorem on the independence of characters

THEOREM 5.14 (DEDEKIND) *Let F be a field and G a group. Every finite set $\{\chi_1,\ldots,\chi_m\}$ of group homomorphisms $G \to F^\times$ is linearly independent over F, i.e.,*

$$a_1\chi_1 + \cdots + a_m\chi_m = 0 \text{ (as a function } G \to F) \implies a_1 = 0,\ldots,a_m = 0.$$

PROOF. We use induction on m. For $m = 1$, the statement is obvious. Assume it for $m - 1$, and suppose that, for some set $\{\chi_1,\ldots,\chi_m\}$ of homomorphisms $G \to F^\times$ and $a_i \in F$,

$$a_1\chi_1(x) + a_2\chi_2(x) + \cdots + a_m\chi_m(x) = 0 \quad \text{for all } x \in G.$$

We have to show that the a_i are zero. As χ_1 and χ_2 are distinct, they will take distinct values on some $g \in G$. On replacing x with gx in the equation, we find that

$$a_1\chi_1(g)\chi_1(x) + a_2\chi_2(g)\chi_2(x) + \cdots + a_m\chi_m(g)\chi_m(x) = 0 \quad \text{for all } x \in G.$$

On multiplying the first equation by $\chi_1(g)$ and subtracting it from the second, we obtain the equation

$$a_2'\chi_2 + \cdots + a_m'\chi_m = 0, \qquad a_i' = a_i(\chi_i(g) - \chi_1(g)).$$

The induction hypothesis shows that $a_i' = 0$ for $i = 2, 3, \ldots, m$. As $\chi_2(g) - \chi_1(g) \neq 0$, this implies that $a_2 = 0$, and so

$$a_1\chi_1 + a_3\chi_3 + \cdots + a_m\chi_m = 0.$$

The induction hypothesis now shows that the remaining a_i are also zero. $\square$

COROLLARY 5.15 *Let F and E be fields, and let $\sigma_1, \ldots, \sigma_m$ be distinct homomorphisms $F \to E$. Then $\sigma_1, \ldots, \sigma_m$ are linearly independent over E.*

PROOF. Apply the theorem to $\chi_i = \sigma_i | F^\times$. $\square$

COROLLARY 5.16 *Let E be a finite separable extension of F of degree m. Let $\alpha_1, \ldots, \alpha_m$ be a basis for E as an F-vector space, and let $\sigma_1, \ldots, \sigma_m$ be distinct F-homomorphisms from E into a field Ω. Then the matrix whose (i, j)th-entry is $\sigma_i\alpha_j$ is invertible.*

PROOF. If not, there exist $c_i \in \Omega$ such that $\sum_{i=1}^m c_i\sigma_i(\alpha_j) = 0$ for all j. But the map $\sum_{i=1}^m c_i\sigma_i : E \to \Omega$ is F-linear, and so this implies that

$$\sum_{i=1}^m c_i\sigma_i(\alpha) = 0$$

for all $\alpha \in E$, which contradicts Corollary 5.15. $\square$

The normal basis theorem

DEFINITION 5.17 Let E be a finite Galois extension of F. A basis for E as an F-vector space is called a ***normal basis*** if it consists of the conjugates of a single element of E.

In other words, a normal basis is one of the form

$$\{\sigma\alpha \mid \sigma \in \mathrm{Gal}(E/F)\}$$

for some $\alpha \in E$.

THEOREM 5.18 (NORMAL BASIS THEOREM) *Every Galois extension has a normal basis.*

The ***group algebra*** FG of a group G is the F-vector space with basis the elements of G endowed with the multiplication extending that of G. Thus an element of FG is a sum $\sum_{\sigma \in G} a_\sigma \sigma$, $a_\sigma \in F$, and

$$\left(\sum_\sigma a_\sigma \sigma\right)\left(\sum_\sigma b_\sigma \sigma\right) = \sum_\sigma \left(\sum_{\sigma_1 \sigma_2 = \sigma} a_{\sigma_1} b_{\sigma_2}\right)\sigma.$$

Every F-linear action of G on an F-vector space V extends uniquely to an action of FG.

Let E/F be a Galois extension with Galois group G. Then E is an FG-module, and Theorem 5.18 says that there exists an element $\alpha \in E$ such that the map

$$\sum_\sigma a_\sigma \sigma \mapsto \sum_\sigma a_\sigma \sigma \alpha \colon FG \to E$$

is an isomorphism of FG-modules, i.e., that E is a free FG-module of rank 1.

We give three proofs of Theorem 5.18. The first assumes that F is infinite and the second that G is cyclic. Since every Galois extension of a finite field is cyclic (4.20), this covers all cases. The third proof applies to both finite and infinite fields, but uses the Krull–Schmidt theorem.

PROOF FOR INFINITE FIELDS

LEMMA 5.19 *Let $f \in F[X_1, \ldots, X_m]$, and let S be an infinite subset of F. If $f(a_1, \ldots, a_m) = 0$ for all $a_1, \ldots, a_m \in S$, then f is the zero polynomial (i.e., $f = 0$ in $F[X_1, \ldots, X_m]$).*

PROOF. We prove this by induction on m. For $m = 1$, the lemma becomes the statement that a nonzero polynomial in one symbol has only finitely many roots (see 1.7). For $m > 1$, write f as a polynomial in X_m with coefficients in $F[X_1, \ldots, X_{m-1}]$, say,

$$f = \sum c_i(X_1, \ldots, X_{m-1}) X_m^i.$$

For any $(m-1)$-tuple $a_1, \ldots, a_{m-1}$ of elements of S,

$$f(a_1, \ldots, a_{m-1}, X_m)$$

is a polynomial in X_m having every element of S as a root. Therefore, each of its coefficients is zero: $c_i(a_1, \ldots, a_{m-1}) = 0$ for all i. Since this holds for all $(a_1, \ldots, a_{m-1})$, the induction hypothesis shows that $c_i(X_1, \ldots, X_{m-1})$ is the zero polynomial. $\qquad\square$

We now prove 5.18 in the case that F is infinite. Number the elements of G as $\sigma_1,\ldots,\sigma_m$ with σ_1 the identity map.

Suppose that $f \in F[X_1,\ldots,X_m]$ has the property,

$$f(\sigma_1\alpha,\ldots,\sigma_m\alpha) = 0 \quad \text{for all } \alpha \in E.$$

Choose a basis $\alpha_1,\ldots,\alpha_m$ for E as an F-vector space, and let $g(Y_1,\ldots,Y_m) \in E[Y_1,\ldots,Y_m]$ be obtained from f by replacing X_j with $\sum_{i=1}^{m} Y_i\sigma_j\alpha_i$. Then, for all $a_1,\ldots,a_m \in F$,

$$\begin{aligned}
g(a_1,\ldots,a_m) &= f(\textstyle\sum_{i=1}^{m} a_i\sigma_1\alpha_i,\ldots,\sum_{i=1}^{m} a_i\sigma_m\alpha_i)\\
&= f(\sigma_1 \textstyle\sum_{i=1}^{m} a_i\alpha_i,\ldots,\sigma_m \sum_{i=1}^{m} a_i\alpha_i)\\
&= 0,
\end{aligned}$$

and so $g = 0$ (here we use that F is infinite). But the matrix $(\sigma_i\alpha_j)$ is invertible (5.16). Since g is obtained from f by an invertible linear change of variables, f can be obtained from g by the inverse linear change of variables. Therefore it also is zero.

Write $X_i = X(\sigma_i)$, and let $A = (X(\sigma_i\sigma_j))$, i.e., A is the $m \times m$ matrix having X_k in the (i,j)th place if $\sigma_i\sigma_j = \sigma_k$. Then $\det(A)$ is a polynomial in $X_1,\ldots,X_m$, say, $\det(A) = h(X_1,\ldots,X_m)$. Clearly, $h(1,0,\ldots,0)$ is the determinant of a matrix having exactly one 1 in each row and each column and its remaining entries 0. Hence the rows of the matrix are a permutation of the rows of the identity matrix, and so its determinant is ± 1. In particular, the polynomial h is not identically zero, and so there exists an $\alpha \in E^\times$ such that $h(\sigma_1\alpha,\ldots,\sigma_m\alpha)$ $(= \det(\sigma_i\sigma_j\alpha))$ is nonzero. We'll show that $\{\sigma_j\alpha\}$ is a normal basis. For this, it suffices to show that the $\sigma_j\alpha$ are linearly independent over F. Suppose that

$$\sum_{j=1}^{m} a_j\sigma_j\alpha = 0$$

for some $a_j \in F$. On applying $\sigma_1,\ldots,\sigma_m$ successively, we obtain a system of m-equations

$$\sum_{j} a_j\sigma_i\sigma_j\alpha = 0$$

in the m "unknowns" a_j. Because this system of equations is nonsingular, the a_j are zero. This completes the proof.

PROOF WHEN G IS CYCLIC.

Assume that G is generated by an element σ_0 of order n. Then $[E:F] = n$. The minimal polynomial of σ_0 regarded as an endomorphism of the F-vector space E is the monic polynomial in $F[X]$ of least degree such that

$P(\sigma_0) = 0$ (as an endomorphism of E). It has the property that it divides every polynomial $Q(X) \in F[X]$ such that $Q(\sigma_0) = 0$. Since $\sigma_0^n = 1$, $P(X)$ divides $X^n - 1$. On the other hand, Dedekind's theorem on the independence of characters (5.14) implies that $1, \sigma_0, \ldots, \sigma_0^{n-1}$ are linearly independent over F, and so $\deg P(X) > n - 1$. We conclude that $P(X) = X^n - 1$. Therefore, as an $F[X]$-module with X acting as σ_0, E is isomorphic to $F[X]/(X^n - 1)$.[4] For any generator α of E as an $F[X]$-module, $\alpha, \sigma_0\alpha, \ldots, \sigma_0\alpha^{n-1}$ is an F-basis for E.

When F is finite, it is possible to replace the use of Dedekind's theorem (5.14) with a counting argument.

UNIFORM PROOF

A module over a ring is **indecomposable** if it is nonzero and cannot be written as a direct sum of two nonzero submodules. The Krull-Schmidt theorem says that every nonzero module M of finite length over a ring can be written as a direct sum of indecomposable modules and that the indecomposable modules occurring in a decomposition are unique up to order and isomorphism. Thus $M = \bigoplus_i m_i M_i$ where M_i is indecomposable and $m_i M_i$ denotes the direct sum of m_i copies of M_i; the set of isomorphism classes of the M_i is uniquely determined and, when we choose the M_i to be pairwise nonisomorphic, each m_i is uniquely determined. From this it follows that two modules M and M' of finite length over a ring are isomorphic if $mM \approx mM'$ for some $m \geq 1$.

Consider the F-vector space $E \otimes_F E$. We let E act on the first factor, and G act on the second factor (so $a(x \otimes y) = ax \otimes y$, $a \in E$, and $\sigma(x \otimes y) = x \otimes \sigma y$, $\sigma \in G$). We'll prove Theorem 5.18 by showing that

$$\underbrace{FG \oplus \cdots \oplus FG}_{n} \approx E \otimes_F E \approx \underbrace{E \oplus \cdots \oplus E}_{n}$$

as FG-modules ($n = [E:F]$).

For $\sigma \in G$, let $\lambda_\sigma \colon E \otimes_F E \to E$ denote the map $x \otimes y \mapsto x \cdot \sigma y$. Then λ_σ is obviously E-linear, and $\lambda_\sigma(\tau z) = \lambda_{\sigma\tau}(z)$ for all $\tau \in G$ and $z \in E \otimes_F E$. I claim that $\{\lambda_\sigma \mid \sigma \in G\}$ is an E-basis for $\operatorname{Hom}_{E\text{-linear}}(E \otimes_F E, E)$. As this

---

[4] This follows from the structure theory of a vector space V equipped with an endomorphism α. As an $F[X]$-module, V is a direct sum $V = V_1 \oplus V_2 \oplus \cdots$ with V_i isomorphic to $F[X]/(P_i(X))$, where P_i is a monic polynomial. The minimal polynomial of α is the lcm of the P_i and its characteristic polynomial is $\prod P_i$. In our case, both polynomials are $X^n - 1$, and so the P_i are relatively prime with product $X^n - 1$. By the Chinese remainder theorem, V is isomorphic to $k[X]/(X^n - 1)$.

space has dimension n, it suffices to show that the set is linearly independent. But if $\sum_\sigma c_\sigma \lambda_\sigma = 0$, $c_\sigma \in E$, then

$$0 = \sum_\sigma c_\sigma(\lambda_\sigma(1 \otimes y)) = \sum_\sigma c_\sigma \cdot \sigma y$$

for all $y \in E$, which implies that all $c_\sigma = 0$ by Dedekind's theorem 5.14.

Consider the map

$$\phi: E \otimes_F E \to EG, \quad z \mapsto \sum_\sigma \lambda_\sigma(z) \cdot \sigma^{-1}.$$

Then ϕ is E-linear. If $\phi(z) = 0$, then $\lambda_\sigma(z) = 0$ for all $\sigma \in G$, and so $z = 0$ in $E \otimes_F E$ (because the λ_σ span the dual space). Therefore ϕ is injective, and as $E \otimes_F E$ and EG both have dimension n over E, it is an isomorphism. For $\tau \in G$,

$$\begin{aligned}
\phi(\tau z) &= \sum_\sigma \lambda_\sigma(\tau z) \cdot \sigma^{-1} \\
&= \sum_\sigma \lambda_{\sigma\tau}(z) \cdot \tau(\sigma\tau)^{-1} \\
&= \tau\phi(z),
\end{aligned}$$

and so ϕ is an isomorphism of EG-modules. Thus

$$E \otimes_F E \simeq EG \approx FG \oplus \cdots \oplus FG$$

as an FG-module.

On the other hand, for any basis $\{e_1, \ldots, e_n\}$ for E as an F-vector space,

$$E \otimes_F E = (e_1 \otimes E) \oplus \cdots \oplus (e_n \otimes E) \simeq E \oplus \cdots \oplus E$$

as FG-modules. This completes the proof.

NOTES The normal basis theorem was stated for finite fields by Eisenstein in 1850, and proved for finite fields by Hensel in 1888. Dedekind used normal bases in number fields in his work on the discriminant in 1880, but he had no general proof. Emmy Noether gave a proof for some infinite fields (1932) and Deuring gave a uniform proof (also 1932). The above uniform proof simplifies that of Deuring — see Blessenohl, *On the normal basis theorem*. Note Mat. 27 (2007), 5–10. According to the Wikipedia, normal bases are frequently used in cryptographic applications.

Hilbert's Theorem 90

Let G be a group. A G-***module*** is an abelian group M together with an ***action*** of G, i.e., a map $G \times M \to M$ such that

(a) $\sigma(m+m') = \sigma m + \sigma m'$ for all $\sigma \in G$, $m, m' \in M$;

(b) $(\sigma\tau)(m) = \sigma(\tau m)$ for all $\sigma, \tau \in G$, $m \in M$;

(c) $1_G m = m$ for all $m \in M$.

Thus, to give an action of G on M is the same as giving a homomorphism $G \to \mathrm{Aut}(M)$.

EXAMPLE 5.20 Let E be a Galois extension of F with Galois group G. Then $(E, +)$ and $(E^\times, \cdot)$ are G-modules.

Let M be a G-module. A ***crossed homomorphism*** is a map $f : G \to M$ such that
$$f(\sigma\tau) = f(\sigma) + \sigma f(\tau) \text{ for all } \sigma, \tau \in G.$$

Note that the condition implies that $f(1) = f(1 \cdot 1) = f(1) + f(1)$, and so $f(1) = 0$.

EXAMPLE 5.21 (a) Let $f : G \to M$ be a crossed homomorphism. For any $\sigma \in G$,

$$f(\sigma^2) = f(\sigma) + \sigma f(\sigma),$$
$$f(\sigma^3) = f(\sigma \cdot \sigma^2) = f(\sigma) + \sigma f(\sigma) + \sigma^2 f(\sigma)$$
$$\cdots$$
$$f(\sigma^n) = f(\sigma) + \sigma f(\sigma) + \cdots + \sigma^{n-1} f(\sigma).$$

Thus, if G is a cyclic group of order n generated by σ, then a crossed homomorphism $f : G \to M$ is determined by its value, x say, on σ, and x satisfies the equation

$$x + \sigma x + \cdots + \sigma^{n-1} x = 0, \tag{10}$$

Moreover, if $x \in M$ satisfies (10), then the formulas

$$f(\sigma^i) = x + \sigma x + \cdots + \sigma^{i-1} x$$

define a crossed homomorphism $f : G \to M$. Thus, for a finite cyclic group $G = \langle \sigma \rangle$, there is a one-to-one correspondence

$$\{\text{crossed homomorphisms } f : G \to M\} \overset{f \leftrightarrow f(\sigma)}{\longleftrightarrow} \{x \in M \text{ satisfying } (10)\}.$$

(b) For every $x \in M$, we obtain a crossed homomorphism by putting

$$f(\sigma) = \sigma x - x, \qquad \text{all } \sigma \in G.$$

Such a crossed homomorphism is said to be **principal**.

(c) If G acts trivially on M, i.e., $\sigma m = m$ for all $\sigma \in G$ and $m \in M$, then a crossed homomorphism is simply a homomorphism, and there are no nonzero principal crossed homomorphisms.

The sum and difference of two crossed homomorphisms is again a crossed homomorphism, and the sum and difference of two principal crossed homomorphisms is again principal. Thus we can define

$$H^1(G, M) = \frac{\{\text{crossed homomorphisms}\}}{\{\text{principal crossed homomorphisms}\}}$$

(quotient abelian group). There are also cohomology groups $H^n(G, M)$ for $n > 1$, but we shall not be concerned with them. An exact sequence of G-modules

$$0 \to M' \to M \to M'' \to 0$$

gives rise to an exact sequence

$$0 \to M'^G \to M^G \to M''^G \xrightarrow{d} H^1(G, M') \to H^1(G, M) \to H^1(G, M'').$$

Let $m'' \in M''^G$, and let $m \in M$ map to m''. For all $\sigma \in G$, $\sigma m - m$ lies in the submodule M' of M, and $\sigma \mapsto \sigma m - m : G \to M'$ is a crossed homomorphism, whose class we define to be $d(m'')$. We leave it as an exercise for the reader to check the exactness.

EXAMPLE 5.22 Let $\pi \colon \tilde{X} \to X$ be the universal covering space of a topological space X, and let Γ be the group of covering transformations. Under some fairly general hypotheses, a Γ-module M will define a sheaf $\mathcal{M}$ on X, and $H^1(X, \mathcal{M}) \simeq H^1(\Gamma, M)$. For example, when $M = \mathbb{Z}$ with the trivial action of Γ, this becomes the isomorphism $H^1(X, \mathbb{Z}) \simeq H^1(\Gamma, \mathbb{Z}) = \text{Hom}(\Gamma, \mathbb{Z})$.

THEOREM 5.23 *Let E be a Galois extension of F with group G; then $H^1(G, E^\times) = 0$, i.e., every crossed homomorphism $G \to E^\times$ is principal.*

PROOF. Let f be a crossed homomorphism $G \to E^\times$. In multiplicative notation, this means that

$$f(\sigma\tau) = f(\sigma) \cdot \sigma(f(\tau)), \quad \sigma, \tau \in G,$$

and we have to find a $\gamma \in E^\times$ such that $f(\sigma) = \frac{\sigma\gamma}{\gamma}$ for all $\sigma \in G$. Because the $f(\tau)$ are nonzero, Corollary 5.15 implies that

$$\sum_{\tau \in G} f(\tau)\tau \colon E \to E$$

is not the zero map, i.e., there exists an $\alpha \in E$ such that

$$\beta \overset{\text{def}}{=} \sum_{\tau \in G} f(\tau)\tau\alpha \neq 0.$$

But then, for $\sigma \in G$,

$$\sigma\beta = \sum_{\tau \in G} \sigma(f(\tau)) \cdot \sigma\tau(\alpha)$$
$$= \sum_{\tau \in G} f(\sigma)^{-1} f(\sigma\tau) \cdot \sigma\tau(\alpha)$$
$$= f(\sigma)^{-1} \sum_{\tau \in G} f(\sigma\tau)\sigma\tau(\alpha),$$

which equals $f(\sigma)^{-1}\beta$ because, as τ runs over G, so also does $\sigma\tau$. Therefore,

$$f(\sigma) = \frac{\beta}{\sigma(\beta)} = \frac{\sigma(\beta^{-1})}{\beta^{-1}}.$$

$\square$

Let E be a Galois extension of F with Galois group G. We define the *norm* of an element $\alpha \in E$ to be

$$\text{Nm}\,\alpha = \prod_{\sigma \in G} \sigma\alpha.$$

For $\tau \in G$,

$$\tau(\text{Nm}\,\alpha) = \prod_{\sigma \in G} \tau\sigma\alpha = \text{Nm}\,\alpha,$$

and so $\text{Nm}\,\alpha \in F$. The map

$$\alpha \mapsto \text{Nm}\,\alpha \colon E^{\times} \to F^{\times}$$

is a obviously a homomorphism.

EXAMPLE 5.24 The norm map $\mathbb{C}^{\times} \to \mathbb{R}^{\times}$ is $\alpha \mapsto |\alpha|^2$ and the norm map $\mathbb{Q}[\sqrt{d}]^{\times} \to \mathbb{Q}^{\times}$ is $a + b\sqrt{d} \mapsto a^2 - db^2$.

We are interested in determining the kernel of the norm map. Clearly an element of the form $\frac{\beta}{\tau\beta}$ has norm 1, and our next result shows that, for cyclic extensions, all elements with norm 1 are of this form.

COROLLARY 5.25 (HILBERT'S THEOREM 90) *Let E be a finite cyclic extension of F, and let σ generate $\text{Gal}(E/F)$. Let $\alpha \in E^{\times}$; if $\text{Nm}_{E/F}\,\alpha = 1$, then $\alpha = \beta/\sigma\beta$ for some $\beta \in E$.*

PROOF. Let $m = [E:F]$. The condition on α is that $\alpha \cdot \sigma\alpha \cdots \sigma^{m-1}\alpha = 1$, and so (see 5.21a) there is a crossed homomorphism $f \colon \langle\sigma\rangle \to E^{\times}$ with $f(\sigma) = \alpha$. Theorem 5.23 now shows that f is principal, which means that there is a β with $f(\sigma) = \beta/\sigma\beta$.

$\square$

Aside 5.26 With the obvious notion of morphism, the G-modules form a category. This is essentially the same as the category of $\mathbb{Z}G$-modules, where $\mathbb{Z}G$ is the group ring of G (Wikipedia: group ring). The category has enough injectives, and the H^1 is the first right derived functor of $M \rightsquigarrow M^G$.

Notes The corollary is Satz 90 in Hilbert, *Theorie der Algebraischen Zahlkörper*, 1897. The theorem was discovered by Kummer in the special case of $\mathbb{Q}[\zeta_p]/\mathbb{Q}$, and generalized to Theorem 5.23 by Emmy Noether. Theorem 5.23, as well as various vast generalizations of it, are also referred to as Hilbert's Theorem 90.

Cyclic extensions

Let F be a field containing a primitive nth root of 1, some $n \geq 2$. Then the group μ_n of nth roots of 1 in F is a cyclic subgroup of $F^\times$ of order n, and we let ζ denote a generator of μ_n. In this section, we classify the cyclic extensions of F of degree n.

Consider a field $E = F[\alpha]$ generated by an element α whose nth power, but no smaller power, lies in F. Then α is a root of $X^n - a$, where $a = \alpha^n$, and the remaining roots are the elements $\zeta^i \alpha$, $1 \leq i \leq n-1$. Since these all lie in E, it is Galois over F, with Galois group G say. For every $\sigma \in G$, $\sigma\alpha$ is also a root of $X^n - a$, and so $\sigma\alpha = \zeta^i \alpha$ for some i. Hence $\sigma\alpha/\alpha \in \mu_n$. The map

$$\sigma \mapsto \sigma\alpha/\alpha \colon G \to \mu_n$$

is unchanged when α is replaced by a conjugate $\beta = \zeta^i \alpha$ (because $\zeta \in F$), and it follows that it is a homomorphism:

$$\frac{\sigma\tau\alpha}{\alpha} = \frac{\sigma(\tau\alpha)}{\tau\alpha}\frac{\tau\alpha}{\alpha}.$$

If σ lies in the kernel of the map $G \to \mu_n$, then $\sigma\alpha = \alpha$, and so σ is the identity map. Thus the homomorphism $G \to \mu_n$ is injective. If it is not surjective, then G maps into a subgroup μ_d of μ_n, some $d \,|\, n$, $d < n$. In this case, $(\sigma\alpha/\alpha)^d = 1$, i.e., $\sigma\alpha^d = \alpha^d$, for all $\sigma \in G$, and so $\alpha^d \in F$, contradicting the hypothesis on α. Thus the map is surjective. We have proved the first part of the following statement.

Proposition 5.27 *Let F be a field containing a primitive nth root of 1. Let $E = F[\alpha]$, where $\alpha^n \in F$ and no smaller power of α is in F. Then E is a Galois extension of F with cyclic Galois group of order n. Conversely, if E is a cyclic extension of F of degree n, then $E = F[\alpha]$ for some α with $\alpha^n \in F$.*

PROOF. It remains to prove the last statement. Let σ generate G and let ζ generate μ_n. It suffices to find an element $\alpha \in E^\times$ such that $\sigma\alpha = \zeta^{-1}\alpha$, for then α^n is the smallest power of α lying in F. As $1,\sigma,\ldots,\sigma^{n-1}$ are distinct homomorphisms $F^\times \to F^\times$, Dedekind's Theorem 5.14 shows that $\sum_{i=0}^{n-1} \zeta^i \sigma^i$ is not the zero function, and so there exists a γ such that $\alpha \stackrel{\text{def}}{=} \sum \zeta^i \sigma^i \gamma \neq 0$. Now $\sigma\alpha = \zeta^{-1}\alpha$. $\qquad\square$

Let F be a field containing a primitive nth root of 1, and let Ω be a field containing F. Let $E = F[\alpha]$, where α is an element of Ω such that $\alpha^n \in F$. Then E (as a subfield of Ω) depends only on $a \stackrel{\text{def}}{=} \alpha^n$, and so we denote it by $F[a^{\frac{1}{n}}]$.

PROPOSITION 5.28 *Let F be a field containing a primitive nth root of 1, and let Ω be a field containing F. Two cyclic extensions $F[a^{\frac{1}{n}}]$ and $F[b^{\frac{1}{n}}]$ of F in Ω of degree n are equal if and only if $a = b^r c^n$ for some $r \in \mathbb{Z}$ relatively prime to n and some $c \in F^\times$, i.e., if and only if a and b generate the same subgroup of $F^\times/F^{\times n}$.*

PROOF. Only the "only if" part requires proof. We are given that $F[\alpha] = F[\beta]$ with $\alpha^n = a$ and $\beta^n = b$. Let σ be the generator of the Galois group. Then $\sigma\alpha = \zeta\alpha$ and $\sigma\beta = \zeta^i\beta$ for some primitive nth root of 1, ζ, and integer i prime to n. We can write

$$\beta = \sum_{j=0}^{n-1} c_j \alpha^j, \quad c_j \in F,$$

and then

$$\sigma\beta = \sum_{j=0}^{n-1} c_j \zeta^j \alpha^j.$$

On comparing this with $\sigma\beta = \zeta^i\beta$, we find that $\zeta^i c_j = \zeta^j c_j$ for all j. Hence $c_j = 0$ for $j \neq i$, and therefore $\beta = c_i \alpha^i$. $\qquad\square$

Let Ω be an algebraically closed field containing F. The propositions show that the cyclic extensions of F in Ω of degree n are classified by the cyclic subgroups of $F^\times/F^{\times n}$ of order n.

ASIDE 5.29 (a) It is not difficult to show that the polynomial $X^n - a$ is irreducible in $F[X]$ if a is not a pth power for any prime p dividing n. When we drop the hypothesis that F contains a primitive nth root of 1, this is still true except that, if $4|n$, we need to add the condition that $a \notin -4F^4$. See Lang, *Algebra*, Springer, 2002, VI, §9, Theorem 9.1, p. 297.

(b) If F has characteristic p (hence has no pth roots of 1 other than 1), then $X^p - X - a$ is irreducible in $F[X]$ unless $a = b^p - b$ for some $b \in F$, and when it is irreducible, its Galois group is cyclic of order p (generated by $\alpha \mapsto \alpha + 1$ where α is a root). Moreover, every cyclic extension of F of degree p is the splitting field of such a polynomial.

Kummer theory

Throughout this section, F is a field and ζ is a primitive nth root of 1 in F. In this section, we classify the extensions of F whose Galois group is abelian of exponent n.

Recall that the **exponent** of a finite group G is the smallest integer $n \geq 1$ such that $\sigma^n = 1$ for all $\sigma \in G$. A finite abelian group of exponent n is isomorphic to a subgroup of $(\mathbb{Z}/n\mathbb{Z})^r$ for some r.

Let E/F be a finite Galois extension with Galois group G. From the exact sequence

$$1 \longrightarrow \mu_n \longrightarrow E^\times \xrightarrow{x \mapsto x^n} E^{\times n} \longrightarrow 1$$

we obtain a cohomology sequence

$$1 \longrightarrow \mu_n \longrightarrow F^\times \xrightarrow{x \mapsto x^n} F^\times \cap E^{\times n} \longrightarrow H^1(G, \mu_n) \longrightarrow 1.$$

The sequence ends with 1 because of Hilbert's Theorem 90. Thus we obtain an isomorphism

$$F^\times \cap E^{\times n} / F^{\times n} \to \mathrm{Hom}(G, \mu_n).$$

This map can be described as follows: let a be an element of $F^\times$ that becomes an nth power in E, say $a = \alpha^n$; then a maps to the homomorphism $\sigma \mapsto \frac{\sigma\alpha}{\alpha}$. If G is abelian of exponent n, then

$$|\mathrm{Hom}(G, \mu_n)| = (G : 1).$$

THEOREM 5.30 *The map*

$$E \mapsto F^\times \cap E^{\times n}$$

defines a one-to-one correspondence between the sets of

 (a) *finite abelian extensions of F of exponent n contained in some fixed algebraic closure Ω of F, and*

(b) *subgroups B of $F^\times$ containing $F^{\times n}$ as a subgroup of finite index.*

The extension corresponding to B is $F[B^{\frac{1}{n}}]$, the smallest subfield of Ω containing F and an nth root of every element of B. If $E \leftrightarrow B$, then $[E:F] = (B:F^{\times n})$.

PROOF. For any finite Galois extension E of F, define $B(E) = F^\times \cap E^{\times n}$. Then $E \supset F[B(E)^{\frac{1}{n}}]$, and for any group B containing $F^{\times n}$ as a subgroup of finite index, $B(F[B^{\frac{1}{n}}]) \supset B$. Therefore,

$$[E:F] \geq [F[B(E)^{\frac{1}{n}}]:F] = (B(F[B(E)^{\frac{1}{n}}]):F^{\times n}) \geq (B(E):F^{\times n}).$$

If E/F is abelian of exponent n, then $[E:F] = (B(E):F^{\times n})$, and so equalities hold throughout: $E = F[B(E)^{\frac{1}{n}}]$.

Next consider a group B containing $F^{\times n}$ as a subgroup of finite index, and let $E = F[B^{\frac{1}{n}}]$. Then E is a composite of the extensions $F[a^{\frac{1}{n}}]$ for a running through a set of generators for $B/F^{\times n}$, and so it is a finite abelian extension of exponent n. Therefore

$$a \mapsto \left(\sigma \mapsto \frac{\sigma a^{1/n}}{a^{1/n}} \right) : B(E)/F^{\times n} \to \operatorname{Hom}(G, \mu_n), \quad G = \operatorname{Gal}(E/F),$$

is an isomorphism. This map sends $B/F^{\times n}$ isomorphically onto the subgroup $\operatorname{Hom}(G/H, \mu_n)$ of $\operatorname{Hom}(G, \mu_n)$ where H consists of the $\sigma \in G$ such that $\sigma a^{1/n}/a^{1/n} = 1$ for all $a \in B$. But such a σ fixes all $a^{1/n}$ for $a \in B$, and therefore is the identity automorphism on $E = F[B^{\frac{1}{n}}]$. This shows that $B(E) = B$, and hence $E \mapsto B(E)$ and $B \mapsto F[B^{\frac{1}{n}}]$ are inverse bijections.$\square$

EXAMPLE 5.31 (a) The theorem says that the abelian extensions of $\mathbb{R}$ of exponent 2 are indexed by the subgroups of $\mathbb{R}^\times/\mathbb{R}^{\times 2} = \{\pm 1\}$. This is certainly true.

(b) The theorem says that the finite abelian extensions of $\mathbb{Q}$ of exponent 2 are indexed by the finite subgroups of $\mathbb{Q}^\times/\mathbb{Q}^{\times 2}$. Modulo squares, every nonzero rational number has a unique representative of the form $\pm p_1 \cdots p_r$ with the p_i prime numbers. Therefore $\mathbb{Q}^\times/\mathbb{Q}^{\times 2}$ is a direct sum of cyclic groups of order 2 indexed by the prime numbers plus ∞. The extension corresponding to the subgroup generated by the primes $p_1, \ldots, p_r$ (and -1) is obtained by adjoining the square roots of $p_1, \ldots, p_r$ (and -1) to $\mathbb{Q}$.

REMARK 5.32 Let E be an abelian extension of F of exponent n, and let

$$B(E) = \{a \in F^\times \mid a \text{ becomes an } n\text{th power in } E\}.$$

There is a perfect pairing

$$(a,\sigma) \mapsto \frac{\sigma a^{1/n}}{a^{1/n}} : \frac{B(E)}{F^{\times n}} \times \mathrm{Gal}(E/F) \to \mu_n.$$

Cf. Exercise 2-1 for the case $n = 2$.

Proof of Galois's solvability theorem

LEMMA 5.33 *Let $f \in F[X]$ be separable, and let F' be a field containing F. Then the Galois group of f as an element of $F'[X]$ is a subgroup of the Galois group of f as an element of $F[X]$.*

PROOF. Let E' be a splitting field for f over F', and let $\alpha_1, \ldots, \alpha_m$ be the roots of $f(X)$ in E'. Then $E = F[\alpha_1, \ldots, \alpha_m]$ is a splitting field of f over F. Every element of $\mathrm{Gal}(E'/F')$ permutes the α_i and so maps E into itself. The map $\sigma \mapsto \sigma|E$ is an injection $\mathrm{Gal}(E'/F') \to \mathrm{Gal}(E/F)$. $\square$

THEOREM 5.34 *Let F be a field of characteristic 0. A polynomial in $F[X]$ is solvable in radicals if and only if its Galois group is solvable.*

PROOF. $\Longleftarrow$: Let $f \in F[X]$ have solvable Galois group G_f. Let $F' = F[\zeta]$ where ζ is a primitive nth root of 1 for some large n — for example, $n = (\deg f)!$ will do. The lemma shows that the Galois group G of f as an element of $F'[X]$ is a subgroup of G_f, and hence is also solvable (GT, 6.6a). This means that there is a sequence of subgroups

$$G = G_0 \supset G_1 \supset \cdots \supset G_m = \{1\}$$

such that each G_i is normal in G_{i-1} and G_{i-1}/G_i is cyclic. Let E be a splitting field of $f(X)$ over F', and let $F_i = E^{G_i}$. We have a sequence of fields

$$F \subset F[\zeta] = F' = F_0 \subset F_1 \subset \cdots \subset F_m = E$$

with F_i cyclic over F_{i-1}. Theorem 5.27 shows that $F_i = F_{i-1}[\alpha_i]$ with $\alpha_i^{[F_i:F_{i-1}]} \in F_{i-1}$, each i, and this shows that $f = 0$ is solvable in radicals.

$\Longrightarrow$: It suffices to show that G_f is a quotient of a solvable group (GT, 6.6a). Hence it suffices to find a solvable extension $\tilde{E}$ of F such that $f(X)$ splits in $\tilde{E}[X]$.

We are given that there exists a tower of fields

$$F = F_0 \subset F_1 \subset \cdots \subset F_m$$

such that

(a) $F_i = F_{i-1}[\alpha_i]$, $\alpha_i^{r_i} \in F_{i-1}$;

(b) F_m contains a splitting field for f.

Let $n = r_1 \cdots r_m$, and let Ω be a field Galois over F and containing (a copy of) F_m and a primitive nth root ζ of 1. For example, choose a primitive element γ for F_m over F (see 5.1), and take Ω to be a splitting field of $g(X)(X^n - 1)$ where $g(X)$ is the minimal polynomial of γ over F. Alternatively, apply 2.15.

Let G be the Galois group of Ω/F, and let $\tilde{E}$ be the Galois closure of $F_m[\zeta]$ in Ω. According to (3.18a), $\tilde{E}$ is the composite of the fields $\sigma F_m[\zeta]$, $\sigma \in G$, and so it is generated over F by the elements

$$\zeta, \alpha_1, \alpha_2, \dots, \alpha_m, \sigma\alpha_1, \dots, \sigma\alpha_m, \sigma'\alpha_1, \dots.$$

We adjoin these elements to F one by one to get a sequence of fields

$$F \subset F[\zeta] \subset F[\zeta, \alpha_1] \subset \cdots \subset F' \subset F'' \subset \cdots \subset \tilde{E}$$

in which each field F'' is obtained from its predecessor F' by adjoining an rth root of an element of F' ($r = r_1, \dots, r_m$, or n). According to (5.8) and (5.27), each of these extensions is abelian (and even cyclic after the first), and so $\tilde{E}/F$ is a solvable extension.   □

ASIDE 5.35 One of Galois's major achievements was to show that an irreducible polynomial of prime degree in $\mathbb{Q}[X]$ is solvable by radicals if and only if its splitting field is generated by any two roots of the polynomial.[5] This theorem of Galois answered a question on mathoverflow in 2010 (mo24081). For a partial generalization of Galois's theorem, see mo110727.

Symmetric polynomials

Let R be a commutative ring (with 1). A polynomial $P \in R[X_1, \dots, X_n]$ is said to be **symmetric** if it is unchanged when the X_i are permuted, i.e., if

$$P(X_{\sigma(1)}, \dots, X_{\sigma(n)}) = P(X_1, \dots, X_n) \quad \text{for all } \sigma \in S_n.$$

---

[5]Pour qu'une équation de degré premier soit résoluble par radicaux, il faut et il suffit que deux quelconques de ces racines étant connues, les autres s'en déduisent rationnellement (Évariste Galois, Bulletin de M. Férussac, XIII (avril 1830), p. 271).

For example

$$p_1 = \sum_i X_i \qquad\qquad = X_1 + X_2 + \cdots + X_n,$$
$$p_2 = \sum_{i<j} X_i X_j \qquad\quad = X_1 X_2 + X_1 X_3 + \cdots + X_1 X_n + X_2 X_3 + \cdots,$$
$$p_3 = \sum_{i<j<k} X_i X_j X_k \quad = X_1 X_2 X_3 + \cdots$$
$$\cdots$$
$$p_r = \sum_{i_1 < \cdots < i_r} X_{i_1} \ldots X_{i_r}$$
$$\cdots$$
$$p_n = X_1 X_2 \cdots X_n$$

are all symmetric because p_r is the sum of *all* monomials of degree r that are products of distinct X_i. These particular polynomials are called the ***elementary symmetric polynomials***.

THEOREM 5.36 (SYMMETRIC POLYNOMIALS THEOREM) *Every symmetric polynomial P in $R[X_1, \ldots, X_n]$ is a polynomial in the elementary symmetric polynomials with coefficients in R, i.e., $P \in R[p_1, \ldots, p_n]$.*

PROOF. We define an ordering on the monomials in the X_i by requiring that

$$X_1^{i_1} X_2^{i_2} \cdots X_n^{i_n} > X_1^{j_1} X_2^{j_2} \cdots X_n^{j_n}$$

if either

$$i_1 + i_2 + \cdots + i_n > j_1 + j_2 + \cdots + j_n$$

or equality holds and, for some s,

$$(i_1, \ldots, i_s) = (j_1, \ldots, j_s) \text{ but } i_{s+1} > j_{s+1}.$$

For example,
$$X_1 X_2 X_3^3 > X_1 X_2^2 X_3 > X_1 X_2 X_3^2.$$

Let $P(X_1, \ldots, X_n)$ be a symmetric polynomial, and let $X_1^{i_1} \cdots X_n^{i_n}$ be the highest monomial occurring in P with a nonzero coefficient, so

$$P = c X_1^{i_1} \cdots X_n^{i_n} + \text{lower terms}, \quad c \neq 0.$$

Because P is symmetric, it contains all monomials obtained from $X_1^{i_1} \cdots X_n^{i_n}$ by permuting the X. Hence $i_1 \geq i_2 \geq \cdots \geq i_n$.

The highest monomial in p_i is $X_1 \cdots X_i$, and it follows that the highest monomial in $p_1^{d_1} \cdots p_n^{d_n}$ is

$$X_1^{d_1+d_2+\cdots+d_n} X_2^{d_2+\cdots+d_n} \cdots X_n^{d_n}. \tag{11}$$

Therefore the highest monomial of

$$P(X_1,\ldots,X_n) - c p_1^{i_1-i_2} p_2^{i_2-i_3} \cdots p_n^{i_n} \tag{12}$$

is strictly less than the highest monomial in $P(X_1,\ldots,X_n)$. We can repeat this argument with the polynomial (12), and after a finite number of steps, we will arrive at a representation of P as a polynomial in $p_1,\ldots,p_n$. $\square$

REMARK 5.37 (a) The proof is algorithmic. Consider, for example,

$$\begin{aligned}
P(X_1, X_2) &= (X_1 + 7X_1 X_2 + X_2)^2 \\
&= X_1^2 + 2X_1 X_2 + 14X_1^2 X_2 + X_2^2 + 14X_1 X_2^2 + 49X_1^2 X_2^2.
\end{aligned}$$

The highest monomial is $49X_1^2 X_2^2$, and so we subtract $49p_2^2$, to get

$$P - 49p_2^2 = X_1^2 + 2X_1 X_2 + 14X_1^2 X_2 + X_2^2 + 14X_1 X_2^2.$$

Continuing, we get

$$P - 49p_2^2 - 14p_1 p_2 = X_1^2 + 2X_1 X_2 + X_2^2$$

and finally,

$$P - 49p_2^2 - 14p_1 p_2 - p_1^2 = 0.$$

(Wikipedia: elementary symmetric polynomials).

(b) The expression of P as a polynomial in the p_i in 5.36 is unique. Otherwise, by subtracting, we would get a nontrivial polynomial $Q(p_1,\ldots,p_n)$ in the p_i which is zero when expressed as a polynomial in the X_i. But the highest monomials (11) in the polynomials $p_1^{d_1} \cdots p_n^{d_n}$ are distinct (the map $(d_1,\ldots,d_n) \mapsto (d_1 + \cdots + d_n,\ldots,d_n)$ is injective), and so they cannot cancel.

Let

$$f(X) = X^n + a_1 X^{n-1} + \cdots + a_n \in R[X],$$

and suppose that f splits over some ring S containing R:

$$f(X) = \prod_{i=1}^n (X - \alpha_i), \quad \alpha_i \in S.$$

Then

$$a_1 = -p_1(\alpha_1,\ldots,\alpha_n),\ a_2 = p_2(\alpha_1,\ldots,\alpha_n),\ \ldots,\ a_n = (-1)^n p_n(\alpha_1,\ldots,\alpha_n).$$

Thus the *elementary* symmetric polynomials in the roots of $f(X)$ lie in R, and so the theorem shows that *every* symmetric polynomial in the roots of $f(X)$ lies in R. For example, the discriminant

$$D(f) \overset{\text{def}}{=} \prod_{i<j} (\alpha_i - \alpha_j)^2$$

of f lies in R.

THEOREM 5.38 (SYMMETRIC FUNCTIONS THEOREM) *Let F be a field. If S_n acts on $F(X_1,\ldots,X_n)$ by permuting the X_i, the field of invariants is $F(p_1,\ldots,p_n)$.*

PROOF. Let $f \in F(X_1,\ldots,X_n)$ be symmetric (i.e., fixed by S_n). Set $f = g/h$, $g,h \in F[X_1,\ldots,X_n]$. The polynomials $H = \prod_{\sigma \in S_n} \sigma h$ and Hf are symmetric, and therefore lie in $F[p_1,\ldots,p_n]$ by Theorem 5.36. Hence their quotient $f = Hf/H$ lies in $F(p_1,\ldots,p_n)$. $\qquad\square$

COROLLARY 5.39 *The field $F(X_1,\ldots,X_n)$ is Galois over $F(p_1,\ldots,p_n)$ with Galois group S_n (acting by permuting the X_i).*

PROOF. We have shown that $F(p_1,\ldots,p_n) = F(X_1,\ldots,X_n)^{S_n}$, and so this follows from (3.10). $\qquad\square$

The field $F(X_1,\ldots,X_n)$ is the splitting field over $F(p_1,\ldots,p_n)$ of

$$g(T) = (T - X_1)\cdots(T - X_n) = X^n - p_1 X^{n-1} + \cdots + (-1)^n p_n.$$

Therefore, the Galois group of $g(T) \in F(p_1,\ldots,p_n)[T]$ is S_n.

NOTES Symmetric polynomials played an important role in the work of Galois. In his *Mémoire sur les conditions de résolubilité des équations par radicaux*, he prove the following proposition:

> Let f be a polynomial with coefficients $\sigma_1,\ldots,\sigma_n$. Let $x_1,\ldots,x_n$ be its roots, and let $U,V,\ldots$ be certain numbers that are rational functions in the x_i. Then there exists a group G of permutations of the x_i such that the rational functions in the x_i that are fixed under all permutations in G are exactly those that are rationally expressible in terms of $\sigma_1,\ldots,\sigma_n$ and $U,V,\ldots$

When we take $U,V,\ldots$ to be the elements of a field E intermediate between the field of coefficients of f and the splitting field of f, this says that the exists a group G of permutations of the x_i whose fixed field (when G acts on the splitting field) is exactly E.

The general polynomial of degree n

When we say that the roots of

$$aX^2 + bX + c$$

are

$$\frac{-b \pm \sqrt{b^2 - 4ac}}{2a}$$

we are thinking of a, b, c as symbols: for any particular values of a, b, c, the formula gives the roots of the particular equation. We'll prove in this section that there is no similar formula for the roots of the "general polynomial" of degree ≥ 5.

We define the ***general polynomial of degree*** n to be

$$f(X) = X^n - t_1 X^{n-1} + \cdots + (-1)^n t_n \in F[t_1, \ldots, t_n][X]$$

where the t_i are symbols. We'll show that, when we regard f as a polynomial in X with coefficients in the field $F(t_1, \ldots, t_n)$, its Galois group is S_n. Then Theorem 5.34 proves the above remark (at least in characteristic zero).

THEOREM 5.40 *The Galois group of the general polynomial of degree n is S_n.*

PROOF. Let $f(X)$ be the general polynomial of degree n,

$$f(X) = X^n - t_1 X^{n-1} + \cdots + (-1)^n t_n \in F[t_1, \ldots, t_n][X].$$

If we can show that the homomorphism

$$t_i \mapsto p_i \colon F[t_1, \ldots, t_n] \to F[p_1, \ldots, p_n]$$

is injective, then it will extend to an isomorphism

$$F(t_1, \ldots, t_n) \to F(p_1, \ldots, p_n)$$

sending $f(X)$ to

$$g(X) = X^n - p_1 X^{n-1} + \cdots + (-1)^n p_n \in F(p_1, \ldots, p_n)[X].$$

Then the statement will follow from Corollary 5.39.

We now prove that the homomorphism is injective.[6] Suppose on the contrary that there exists a $P(t_1,\ldots,t_n)$ such that $P(p_1,\ldots,p_n) = 0$. Equation (11), p. 101, shows that if $m_1(t_1,\ldots,t_n)$ and $m_2(t_1,\ldots,t_n)$ are distinct monomials, then $m_1(p_1,\ldots,p_n)$ and $m_2(p_1,\ldots,p_n)$ have distinct highest monomials. Therefore, cancellation cannot occur, and so $P(t_1,\ldots,t_n)$ must be the zero polynomial.   □

ASIDE 5.41 Since S_n occurs as a Galois group over $\mathbb{Q}$, and every finite group occurs as a subgroup of some S_n, it follows that every finite group occurs as a Galois group over some finite extension of $\mathbb{Q}$, but does every finite group occur as a Galois group over $\mathbb{Q}$ itself? In other words, does every finite group occur as the Galois group of some $f \in \mathbb{Q}[X]$? This is known as the **inverse Galois problem**, which is still open.

The Hilbert-Noether program for proving this was the following. Hilbert proved that if G occurs as the Galois group of an extension $E \supset \mathbb{Q}(t_1,\ldots,t_n)$ (the t_i are symbols), then it occurs infinitely often as a Galois group over $\mathbb{Q}$. For the proof, realize E as the splitting field of a polynomial $f(X) \in k[t_1,\ldots,t_n][X]$ and prove that for infinitely many values of the t_i, the polynomial you obtain in $\mathbb{Q}[X]$ has Galois group G. Emmy Noether conjectured the following: Let $G \subset S_n$ act on $F(X_1,\ldots,X_n)$ by permuting the X_i; then $F(X_1,\ldots,X_n)^G \approx F(t_1,\ldots,t_n)$ (for symbols t_i). However, Swan proved in 1969 that the conjecture is false for G the cyclic group of order 47. Hence this approach cannot lead to a proof that all finite groups occur as Galois groups over $\mathbb{Q}$, but it does not exclude other approaches. For more information on the problem, see Serre, *Lectures on the Mordell-Weil Theorem*, 1989, Chapters 9, 10; Serre, *Topics in Galois Theory*, 1992; and Wikipedia: inverse Galois problem.

ASIDE 5.42 Take $F = \mathbb{C}$, and consider the subset of $\mathbb{C}^{n+1}$ defined by the equation

$$X^n - T_1 X^{n-1} + \cdots + (-1)^n T_n = 0.$$

It is a beautiful complex manifold S of dimension n. Consider the projection

$$\pi : S \to \mathbb{C}^n, \quad (x, t_1, \ldots, t_n) \mapsto (t_1, \ldots, t_n).$$

Its fibre over a point $(a_1, \ldots, a_n)$ is the set of roots of the polynomial

$$X^n - a_1 X^{n-1} + \cdots + (-1)^n a_n.$$

The discriminant $D(f)$ of $f(X) = X^n - T_1 X^{n-1} + \cdots + (-1)^n T_n$ is a polynomial in $\mathbb{C}[T_1,\ldots,T_n]$. Let Δ be the zero set of $D(f)$ in $\mathbb{C}^n$. Then over each point of $\mathbb{C}^n \smallsetminus \Delta$, there are exactly n points of S, and $S \smallsetminus \pi^{-1}(\Delta)$ is a covering space over $\mathbb{C}^n \smallsetminus \Delta$.

---

[6]To say that the homomorphism is injective means that the p_i are algebraically independent over F (see p. 147). This can be proved by noting that, because $F(X_1,\ldots,X_n)$ is algebraic over $F(p_1,\ldots,p_n)$, the latter must have transcendence degree n (see §8).

NOTES As far back as 1500 BCE, the Babylonians (at least) knew a general formula for the roots of a quadratic polynomial. Cardan (about 1515 CE) found a general formula for the roots of a cubic polynomial. Ferrari (about 1545) found a general formula for the roots of a quartic polynomial (he introduced the resolvent cubic, and used Cardan's result). Over the next 275 years there were many fruitless attempts to obtain similar formulas for higher degree polynomials until (about 1820) Ruffini and Abel proved that there are none.

Norms and traces

Recall that, for an $n \times n$ matrix $A = (a_{ij})$

$$\mathrm{Tr}(A) = \sum_i a_{ii} \qquad\qquad\qquad \text{trace of } A$$
$$\det(A) = \sum_{\sigma \in S_n} \mathrm{sign}(\sigma) a_{1\sigma(1)} \cdots a_{n\sigma(n)}, \quad \text{determinant of } A$$
$$c_A(X) = \det(X I_n - A) \qquad\qquad \text{characteristic polynomial of } A.$$

Moreover,

$$c_A(X) = X^n - \mathrm{Tr}(A) X^{n-1} + \cdots + (-1)^n \det(A).$$

None of these is changed when A is replaced by its conjugate UAU^{-1} by an invertible matrix U. Therefore, for any endomorphism α of a finite-dimensional vector space V, we can define

$$\mathrm{Tr}(\alpha) = \mathrm{Tr}(A), \quad \det(\alpha) = \det(A), \quad c_\alpha(X) = c_A(X),$$

where A is the matrix of α with respect to a basis of V. If β is a second endomorphism of V,

$$\mathrm{Tr}(\alpha + \beta) = \mathrm{Tr}(\alpha) + \mathrm{Tr}(\beta);$$
$$\det(\alpha\beta) = \det(\alpha) \det(\beta).$$

The coefficients of the characteristic polynomial, $c_\alpha(X) = X^n + c_1 X^{n-1} + \cdots + c_n$, of α have the following description: $c_i = (-1)^i \, \mathrm{Tr}(\alpha | \bigwedge^i V)$ (Bourbaki, *Algèbre*, Chap. III, §8.11).

Now let E be a finite field extension of F of degree n. An element α of E defines an F-linear map

$$\alpha_L : E \to E, \quad x \mapsto \alpha x,$$

and we define

$$\mathrm{Tr}_{E/F}(\alpha) = \mathrm{Tr}(\alpha_L) \qquad \text{(trace of } \alpha\text{)}$$

$$\mathrm{Nm}_{E/F}(\alpha) = \det(\alpha_L) \qquad \text{(norm of } \alpha\text{)}$$

$$c_{\alpha,E/F}(X) = c_{\alpha_L}(X) \qquad \text{(characteristic polynomial of } \alpha\text{)}.$$

Thus, $\mathrm{Tr}_{E/F}$ is a homomorphism $(E,+) \to (F,+)$, and $\mathrm{Nm}_{E/F}$ is a homomorphism $(E^\times,\cdot) \to (F^\times,\cdot)$.

EXAMPLE 5.43 (a) Consider the field extension $\mathbb{C} \supset \mathbb{R}$. For $\alpha = a + bi$, the matrix of α_L with respect to the basis $\{1,i\}$ is $\left(\begin{smallmatrix} a & -b \\ b & a \end{smallmatrix}\right)$, and so

$$\mathrm{Tr}_{\mathbb{C}/\mathbb{R}}(\alpha) = 2a = 2\Re(\alpha),$$

$$\mathrm{Nm}_{\mathbb{C}/\mathbb{R}}(\alpha) = a^2 + b^2 = |\alpha|^2.$$

(b) For $a \in F$, a_L is multiplication by the scalar a. Therefore

$$\mathrm{Tr}_{E/F}(a) = na \quad \mathrm{Nm}_{E/F}(a) = a^n \quad c_{a,E/F}(X) = (X-a)^n$$

where $n = [E:F]$.

Let $E = \mathbb{Q}[\alpha,i]$ be the splitting field of $X^8 - 2$ (see Exercise 4-3). Then E has degree 16 over $\mathbb{Q}$, and so to compute the trace and norm an element of E, the definition requires us to compute the trace and norm of a 16×16 matrix. The next proposition gives us a quicker method.

PROPOSITION 5.44 *Let E/F be a finite extension of fields, and let $f(X)$ be the minimal polynomial of $\alpha \in E$. Then*

$$c_{\alpha,E/F}(X) = f(X)^{[E:F[\alpha]]}.$$

PROOF. Suppose first that $E = F[\alpha]$. In this case, we have to show that $c_\alpha(X) = f(X)$. Note that $\alpha \mapsto \alpha_L$ is an *injective* homomorphism from E into the ring of endomorphisms of E as a vector space over F. The Cayley-Hamilton theorem shows that $c_\alpha(\alpha_L) = 0$, and therefore $c_\alpha(\alpha) = 0$. Hence $f \mid c_\alpha$, but they are monic of the same degree, and so they are equal.

For the general case, let $\beta_1,...,\beta_n$ be a basis for $F[\alpha]$ over F, and let $\gamma_1,...,\gamma_m$ be a basis for E over $F[\alpha]$. As we saw in the proof of (1.20), $\{\beta_i \gamma_k\}$ is a basis for E over F. Write $\alpha\beta_i = \sum a_{ji}\beta_j$. Then, according to the first case proved, $A \overset{\text{def}}{=} (a_{ij})$ has characteristic polynomial $f(X)$. But $\alpha\beta_i\gamma_k = \sum a_{ji}\beta_j\gamma_k$, and so the matrix of α_L with respect to $\{\beta_i\gamma_k\}$ breaks up into $n \times n$ blocks with copies of A down the diagonal and zero matrices elsewhere, from which it follows that $c_{\alpha_L}(X) = c_A(X)^m = f(X)^m$.   □

COROLLARY 5.45 *Suppose that the roots of the minimal polynomial of α are $\alpha_1,\ldots,\alpha_n$ (in some splitting field containing E), and that $[E:F[\alpha]] = m$. Then*

$$\mathrm{Tr}(\alpha) = m\sum_{i=1}^{n}\alpha_i, \qquad \mathrm{Nm}_{E/F}\,\alpha = \left(\prod_{i=1}^{n}\alpha_i\right)^{m}.$$

PROOF. Write the minimal polynomial of α as

$$f(X) = X^n + a_1 X^{n-1} + \cdots + a_n = \prod(X - \alpha_i),$$

so that

$$a_1 = -\sum\alpha_i, \text{ and}$$
$$a_n = (-1)^n \prod\alpha_i.$$

Then

$$c_\alpha(X) = (f(X))^m = X^{mn} + ma_1 X^{mn-1} + \cdots + a_n^m,$$

so that

$$\mathrm{Tr}_{E/F}(\alpha) = -ma_1 = m\sum\alpha_i, \text{ and}$$
$$\mathrm{Nm}_{E/F}(\alpha) = (-1)^{mn}a_n^m = \left(\prod\alpha_i\right)^m. \qquad \square$$

EXAMPLE 5.46 (a) Consider the extension $\mathbb{C} \supset \mathbb{R}$. If $\alpha \in \mathbb{C} \smallsetminus \mathbb{R}$, then

$$c_\alpha(X) = f(X) = X^2 - 2\Re(\alpha)X + |\alpha|^2.$$

If $\alpha \in \mathbb{R}$, then $c_\alpha(X) = (X - a)^2$.

(b) Let E be the splitting field of $X^8 - 2$. Then E has degree 16 over $\mathbb{Q}$ and is generated by $\alpha = \sqrt[8]{2}$ and $i = \sqrt{-1}$ (see Exercise 4-3). The minimal polynomial of α is $X^8 - 2$, and so

$$c_{\alpha,\mathbb{Q}[\alpha]/\mathbb{Q}}(X) = X^8 - 2, \qquad c_{\alpha,E/\mathbb{Q}}(X) = (X^8 - 2)^2$$
$$\mathrm{Tr}_{\mathbb{Q}[\alpha]/\mathbb{Q}}\,\alpha = 0, \qquad\qquad \mathrm{Tr}_{E/\mathbb{Q}}\,\alpha = 0$$
$$\mathrm{Nm}_{\mathbb{Q}[\alpha]/\mathbb{Q}}\,\alpha = -2, \qquad\qquad \mathrm{Nm}_{E/\mathbb{Q}}\,\alpha = 4$$

REMARK 5.47 Let E be a finite extension of F, let Ω be an algebraic closure of F, and let Σ be the set of F-homomorphisms of E into Ω.

When E/F is separable,

$$\mathrm{Tr}_{E/F}\,\alpha = \sum_{\sigma\in\Sigma}\sigma\alpha$$
$$\mathrm{Nm}_{E/F}\,\alpha = \prod_{\sigma\in\Sigma}\sigma\alpha.$$

When $E = F[\alpha]$, this follows from 5.45 and the observation (cf. 2.1b) that the $\sigma\alpha$ are the roots of the minimal polynomial $f(X)$ of α over F. In the general case, the $\sigma\alpha$ are still roots of $f(X)$ in Ω, but now each root of $f(X)$ occurs $[E:F[\alpha]]$ times (because each F-homomorphism $F[\alpha] \to \Omega$ has $[E:F[\alpha]]$ extensions to E). For example, if E is Galois over F with Galois group G, then

$$\mathrm{Tr}_{E/F}\,\alpha = \sum_{\sigma \in G} \sigma\alpha$$
$$\mathrm{Nm}_{E/F}\,\alpha = \prod_{\sigma \in G} \sigma\alpha$$

(in agreement with the previous definition for Galois extensions, p. 93).

In the general case,

$$\mathrm{Tr}_{E/F}\,\alpha = p^e \cdot \sum_{\sigma \in \Sigma} \sigma\alpha$$
$$\mathrm{Nm}_{E/F}\,\alpha = \left(\prod_{\sigma \in \Sigma} \sigma\alpha\right)^{p^e},$$

where p is the characteristic exponent of F and p^e is the degree of E over the separable closure (3.15)(p. 116) of F in E (Bourbaki, *Algèbre*, Chap. V, §8).

PROPOSITION 5.48 *For finite extensions $E \supset M \supset F$, we have*

$$\mathrm{Tr}_{M/F} \circ \mathrm{Tr}_{E/M} = \mathrm{Tr}_{E/F},$$
$$\mathrm{Nm}_{M/F} \circ \mathrm{Nm}_{E/M} = \mathrm{Nm}_{E/F}.$$

PROOF. When E is separable over F, this follows easily from the descriptions in the above remark. We leave the general case as an exercise.   □

PROPOSITION 5.49 *Let $f(X)$ be a monic irreducible polynomial with coefficients in F, and let α be a root of f in some splitting field of f. Then*

$$\mathrm{disc}\,f(X) = (-1)^{m(m-1)/2} \mathrm{Nm}_{F[\alpha]/F}\,f'(\alpha)$$

where f' is the formal derivative $\frac{df}{dX}$ of f.

PROOF. Let $f(X) = \prod_{i=1}^{m}(X - \alpha_i)$ be the factorization of f in the given splitting field, and number the roots so that $\alpha = \alpha_1$. Compute that

$$\mathrm{disc}\,f(X) \overset{\text{def}}{=} \prod_{i<j}(\alpha_i - \alpha_j)^2 = (-1)^{m(m-1)/2} \cdot \prod_{i}\left(\prod_{j \neq i}(\alpha_i - \alpha_j)\right)$$
$$= (-1)^{m(m-1)/2} \cdot \prod_{i} f'(\alpha_i)$$
$$= (-1)^{m(m-1)/2} \mathrm{Nm}_{F[\alpha]/F}(f'(\alpha)) \quad (5.47). \square$$

EXAMPLE 5.50 We compute the discriminant of

$$f(X) = X^n + aX + b, \quad a, b \in F,$$

assumed to be irreducible and separable, by computing the norm of

$$\gamma \stackrel{\text{def}}{=} f'(\alpha) = n\alpha^{n-1} + a, \quad f(\alpha) = 0.$$

On multiplying the equation

$$\alpha^n + a\alpha + b = 0$$

by $n\alpha^{-1}$ and rearranging, we obtain the equation

$$n\alpha^{n-1} = -na - nb\alpha^{-1}.$$

Hence
$$\gamma = n\alpha^{n-1} + a = -(n-1)a - nb\alpha^{-1}.$$

Solving for α gives
$$\alpha = \frac{-nb}{\gamma + (n-1)a}.$$

From the last two equations, it is clear that $F[\alpha] = F[\gamma]$, and so the minimal polynomial of γ over F has degree n also. If we write

$$f\left(\frac{-nb}{X + (n-1)a}\right) = \frac{P(X)}{Q(X)},$$

where

$$P(X) = (X + (n-1)a)^n - na(X + (n-1)a)^{n-1} + (-1)^n n^n b^{n-1}$$
$$Q(X) = (X + (n-1)a)^n / b,$$

then
$$P(\gamma) = f(\alpha) \cdot Q(\gamma) = 0.$$

As
$$Q(\gamma) = \frac{(\gamma + (n-1)a)^n}{b} = \frac{(-nb)^n}{\alpha^n b} \neq 0$$

and $P(X)$ is monic of degree n, it must be the minimal polynomial of γ. Therefore $\operatorname{Nm}\gamma$ is $(-1)^n$ times the constant term of $P(X)$, namely,

$$\operatorname{Nm}\gamma = n^n b^{n-1} + (-1)^{n-1}(n-1)^{n-1} a^n.$$

Therefore,

$$\text{disc}(X^n + aX + b) = (-1)^{n(n-1)/2}(n^n b^{n-1} + (-1)^{n-1}(n-1)^{n-1}a^n),$$

in agreement with 4.39. For example,

$$\text{disc}(X^5 + aX + b) = 5^5 b^4 + 4^4 a^5.$$

This is something PARI does not know how to do (because it does not understand symbols as exponents).

Exercises

5-1 For $a \in \mathbb{Q}$, let G_a be the Galois group of $X^4 + X^3 + X^2 + X + a$. Find integers a_1, a_2, a_3, a_4 such that $i \neq j \implies G_{a_i}$ is not isomorphic to G_{a_j}.

5-2 Prove that the rational solutions $a, b \in \mathbb{Q}$ of Pythagoras's equation $a^2 + b^2 = 1$ are of the form

$$a = \frac{s^2 - t^2}{s^2 + t^2}, \quad b = \frac{2st}{s^2 + t^2}, \qquad s, t \in \mathbb{Q},$$

and deduce that every right triangle with integer sides has sides of length

$$d(m^2 - n^2, 2mn, m^2 + n^2)$$

for some integers d, m, and n (Hint: Apply Hilbert's Theorem 90 to the extension $\mathbb{Q}[i]/\mathbb{Q}$.)

5-3 Prove that a finite extension of $\mathbb{Q}$ can contain only finitely many roots of 1.

5-4 Let E be the splitting field of an irreducible separable polynomial $f \in F[X]$. If no root of f generates E, show that $\text{Gal}(E/F)$ contains a nonnormal subgroup. (Weintraub, Amer. Math. Monthly 128 (2021), no. 8, 753–754.)

Chapter **6**

Algebraic Closures

In this chapter, we use Zorn's lemma to show that every field F has an algebraic closure Ω. Recall that if F is a subfield $\mathbb{C}$, then the algebraic closure of F in $\mathbb{C}$ is an algebraic closure of F (1.47). If F is countable, then the existence of Ω can be proved as in the finite field case (4.24), namely, the set of monic irreducible polynomials in $F[X]$ is countable, and so we can list them $f_1, f_2, \ldots$; define E_i inductively by, $E_0 = F$, $E_i =$ a splitting field of f_i over E_{i-1}; then $\Omega = \bigcup E_i$ is an algebraic closure of F.

The difficulty in showing the existence of an algebraic closure of an arbitrary field F is in the set theory. Roughly speaking, we would like to take a union of a family of splitting fields indexed by the monic irreducible polynomials in $F[X]$, but we need to find a way of doing this that is allowed by the axioms of set theory. After reviewing the statement of Zorn's lemma, we sketch three solutions to the problem.[1]

Zorn's lemma

Definition 6.1 (a) A relation $\leq$ on a set S is a ***partial order*** if it reflexive, transitive, and anti-symmetric ($a \leq b$ and $b \leq a \implies a = b$).

(b) A partial order is a ***total order*** if, for all $s, t \in T$, either $s \leq t$ or $t \leq s$.

(c) An ***upper bound*** for a subset T of a partially ordered set $(S, \leq)$ is an element $s \in S$ such that $t \leq s$ for all $t \in T$.

(d) A ***maximal element*** of a partially ordered set S is an element s such that $s \leq s' \implies s = s'$.

---

[1]There do exist naturally occurring uncountable fields not contained in $\mathbb{C}$. For example, the field of formal Laurent series $F((T))$ over a field F is uncountable even when F is finite.

111

A partially ordered set need not have any maximal elements, for example, the set of finite subsets of an infinite set is partially ordered by inclusion, but it has no maximal element.

ZORN'S LEMMA *Let $(S, \leq)$ be a nonempty partially ordered set for which every totally ordered subset has an upper bound in S. Then S has a maximal element.*

Zorn's lemma is equivalent to the Axiom of Choice, and hence is independent of the axioms of Zermelo-Fraenkel set theory.

The next proposition is a typical application of Zorn's lemma.

PROPOSITION 6.3 *Every nonzero commutative ring A has a maximal ideal (meaning, maximal among* proper *ideals).*

PROOF. Let S be the set of all proper ideals in A, partially ordered by inclusion. If T is a totally ordered set of ideals, then $J = \bigcup_{I \in T} I$ is again an ideal because every finite set of elements of it is contained in a common $I \in T$. It is proper because if $1 \in J$ then $1 \in I$ for some I in T, and I would not be proper. Thus J is an upper bound for T. Now Zorn's lemma implies that S has a maximal element, which is a maximal ideal in A.   □

REMARK 6.4 Zorn's lemma is, in fact, equivalent to the existence of maximal ideals in commutative rings. A weaker axiom, namely, the ultrafilter principle (every filter is contained in a maximal filter) is equivalent to the existence of prime ideals in commutative rings[2] and to the compactness of products of compact spaces,[3] and it implies the axiom of choice for finite sets. This is all we shall need in this chapter.

A condition to be an algebraic closure

PROPOSITION 6.5 *Let Ω/F be an extension of fields. If Ω is algebraic over F and every nonconstant polynomial in $F[X]$ has a root in Ω, then Ω is algebraically closed (hence an algebraic closure of F).*

PROOF. It suffices to show that every monic irreducible polynomial f in $F[X]$ splits in $\Omega[X]$ (see 1.45). Suppose first that f is separable, and let E be a splitting field for f. According to Theorem 5.1, $E = F[\gamma]$ for some $\gamma \in E$. Let $g(X)$ be the minimal polynomial of γ over F. Then $g(X)$ has

---

[2]Rav, Math. Nachr. 79 (1977), 145–165, Cor. 4.4.

[3]Wikipedia: Tychonoff's theorem. Recall that we require compact spaces to be Hausdorff.

coefficients in F, and so it has a root β in Ω. Both of $F[\gamma]$ and $F[\beta]$ are stem fields for g, and so there is an F-isomorphism $F[\gamma] \to F[\beta] \subset \Omega$. As f splits over $F[\gamma]$, it splits over Ω.

This completes the proof when F is perfect. Otherwise, F has characteristic $p \neq 0$, and we let F' denote the set of elements x of Ω such that $x^{p^m} \in F$ for some $m \geq 1$. It is easy to check that F' is a field, and we'll complete the proof of the proposition by showing (a) that F' is perfect, and (b) that every polynomial in $F'[X]$ has a root in Ω.

PROOF OF (a). Let $a \in F'$, so that $b \stackrel{\text{def}}{=} a^{p^m} \in F$ for some m. The polynomial $X^{p^{m+1}} - b$ has coefficients in F, and so it has a root $\alpha \in \Omega$, which automatically lies in F'. Now $\alpha^{p^{m+1}} = a^{p^m}$, which implies that $\alpha^p = a$, because the pth power map is injective on fields of characteristic p. We have shown that F' is perfect.

PROOF OF (b). We first show that Ω is perfect. Suppose that $a \in \Omega$ is not a pth power, and form $\Omega[\alpha]$, where $\alpha^p = a$. Consider $F'[\alpha] \stackrel{p}{\supset} F'[a] \supset F'$. If $g(X)$ is the minimal polynomial of a over F', then $g(X^p)$ is the minimal polynomial of α over F' (it is monic of least degree having α as a zero). In particular it is irreducible, but it is not separable, which contradicts the perfectness of F'.

Let $f(X) \in F'[X]$, say, $f(X) = \sum_i a_i X^i$, $a_i \in F'$. For some m, the polynomial $\sum_i a_i^{p^m} X^i$ has coefficients in F, and therefore has a root $\alpha \in \Omega$. As Ω is perfect, we can write $\alpha = \beta^{p^m}$ with $\beta \in \Omega$. Now

$$(f(\beta))^{p^m} = \left(\sum_i a_i \beta^i\right)^{p^m} = \sum_i a_i^{p^m} \alpha^i = 0,$$

and so β is a root of f. $\qquad\square$

First proof of the existence of algebraic closures

(Bourbaki, *Algèbre*, Chap. V, §4.) Let $(A_i)_{i \in I}$ be a family of commutative algebras over a field F. Define $\bigotimes_F A_i$ to be the quotient of the F-vector space with basis $\prod_{i \in I} A_i$ by the subspace generated by elements of the form:

$(x_i) + (y_i) - (z_i)$ with $x_j + y_j = z_j$ for one $j \in I$ and $x_i = y_i = z_i$ for all $i \neq j$;

$(x_i) - a(y_i)$ with $x_j = ay_j$ for one $j \in I$ and $x_i = y_i$ for all $i \neq j$, (ibid., Chap. II, 3.9). It can be made into a commutative F-algebra in an obvious fashion, and there are canonical homomorphisms $A_i \to \bigotimes_F A_i$ of F-algebras.

For each monic irreducible polynomial $f \in F[X]$, let E_f be the stem field $F[x_f] = F[X_f]/(f)$, and let $\Omega = (\bigotimes_F E_f)/P$, where P is a prime ideal in $\bigotimes_F E_f$ (whose existence is ensured by the ultrafilter principle). Then Ω is an integral domain generated as an F-algebra by elements algebraic over F, and so it is a field algebraic over F (1.32). The composite of the F-homomorphisms $E_f \to \bigotimes_F E_f \to \Omega$, being a homomorphism of fields, is injective. As f has a root in E_f, it has a root in Ω. Hence Ω is an algebraic closure of F by Proposition 6.5.

Second proof of the existence of algebraic closures

(Jacobson, *Lectures in Abstract Algebra*, 1964, Vol. III, p. 144.) After 4.24 we may assume F to be infinite. This implies that the cardinality of every field algebraic over F is the same as that of F (cf. the proof of 1.33). Choose an uncountable set Ξ of cardinality greater than that of F, and identify F with a subset of Ξ. Let S be the set of triples $(E, +, \cdot)$ with $E \subset \Xi$ and $(+, \cdot)$ a field structure on E such that $(E, +, \cdot)$ contains F as a subfield and is algebraic over it. Write $(E, +, \cdot) \leq (E', +', \cdot')$ if the first is a subfield of the second. Apply Zorn's lemma to show that S has maximal elements, and then show that a maximal element is algebraically closed.

Third proof of the existence of algebraic closures

(Emil Artin.) Consider the polynomial ring $F[\ldots, X_f, \ldots]$ in a family of symbols X_f indexed by the monic irreducible polynomials $f \in F[X]$. Let I be the ideal of $F[\ldots, X_f, \ldots]$ generated by the polynomials $f(X_f)$. If $1 \in I$, then

$$g_1 f_1(X_{f_1}) + \cdots + g_n f_n(X_{f_n}) = 1 \qquad (\text{in } F[\ldots, X_f, \ldots])$$

for some $g_i \in F[\ldots, X_f, \ldots]$ and some monic irreducible $f_i \in F[X]$. Let E be an extension of F such that each f_i has a root α_i in E. Under the F-algebra homomorphism $F[\ldots, X_f, \ldots] \to E$ sending

$$\begin{cases} X_{f_i} \mapsto \alpha_i, & i = 1, \ldots, n, \\ X_f \mapsto 0, & f \notin \{f_1, \ldots, f_n\} \end{cases}$$

the above relation becomes $0 = 1$. From this contradiction, we deduce that $1 \notin I$. Let $\Omega = F[\ldots, X_f, \ldots]/P$, where P is a prime ideal containing I (whose existence is ensured by the ultrafilter principle). Then Ω is an integral domain generated as an F-algebra by elements algebraic over F, and so it is a field algebraic over F (1.32). Every monic irreducible $f \in F[X]$ has a root in Ω, and so Ω is an algebraic closure of F by Proposition 6.5.

Any two algebraic closures are isomorphic

THEOREM 6.6 *Let Ω be an algebraic closure of F and E an algebraic extension of F. There exists an F-homomorphism $E \to \Omega$, and, if E is also an algebraic closure of F, then every such homomorphism is an isomorphism.*

PROOF. Suppose first that E is countably generated over F, i.e.,

$$E = F[\alpha_1, ..., \alpha_n, ...].$$

Then we can extend the inclusion map $F \to \Omega$ to $F[\alpha_1]$ (map α_1 to any root of its minimal polynomial in Ω), then to $F[\alpha_1, \alpha_2]$, and so on (see 2.4).

In the uncountable case, we use Zorn's lemma. Let S be the set of pairs (M, φ_M) with M a field $F \subset M \subset E$ and φ_M an F-homomorphism $M \to \Omega$. Write $(M, \varphi_M) \le (N, \varphi_N)$ if $M \subset N$ and $\varphi_N | M = \varphi_M$. This makes S into a partially ordered set. Let T be a totally ordered subset of S. Then $M' = \bigcup_{M \in T} M$ is a subfield of E, and we can define a homomorphism $\varphi': M' \to \Omega$ by requiring that $\varphi'(x) = \varphi_M(x)$ if $x \in M$. The pair (M', φ') is an upper bound for T in S. Hence Zorn's lemma provides us with a maximal element (M, φ) in S. Suppose that $M \ne E$. Then there exists an element $\alpha \in E, \alpha \notin M$. Since α is algebraic over M, we can apply (2.4) to extend φ to $M[\alpha]$, contradicting the maximality of M. Hence $M = E$, and the proof of the first statement is complete.

If E is algebraically closed, then every polynomial $f \in F[X]$ splits in $E[X]$ and hence in $\varphi(E)[X]$. Let $\alpha \in \Omega$, and let $f(X)$ be the minimal polynomial of α. Then $X - \alpha$ is a factor of $f(X)$ in $\Omega[X]$, but, as we just observed, $f(X)$ splits in $\varphi(E)[X]$. Because of unique factorization, this implies that $\alpha \in \varphi(E)$. $\qquad\qquad\square$

The above proof is a typical application of Zorn's lemma: once we know how to do something in a finite (or countable) situation, Zorn's lemma allows us to do it in general.

REMARK 6.7 The above proof used Zorn's lemma. Here is a proof using only the ultrafilter principle. Let Ω and Ω' be algebraic closures of F. For each monic $f \in F[X]$, let H_f be the set of F-isomorphisms from the splitting field of f in Ω to the splitting field of f in Ω'. Then H_f is finite and nonempty, and if $f|g$, then the restriction map $H_g \to H_f$ is surjective.

Let $H = \prod H_f$, and, for each pair (g,h) with $g|h$, let $H_{g,h} = \{(h_f) \in H \mid h_h$ restricts to $h_g\}$. When we give H the product topology (discrete topology on each H_f), it is compact and nonempty (ultrafilter principle). Each subset $H_{g,h}$ is closed because it is the subset of H on which the two obvious maps $H \to H_g$ agree (one map is the projection to H_g and the other passes through the projection to H_h). The sets $H_{g,h}$ have the finite intersection property, and so $\bigcap H_{g,h}$ is nonempty, but any element of $\bigcap H_{g,h}$ defines an isomorphism $\Omega \to \Omega'$.

WARNING 6.8 Even for a finite field F, there will exist uncountably many isomorphisms from one algebraic closure to a second, none of which is to be preferred over any other. Thus it is (uncountably) sloppy to say that the algebraic closure of F is unique. All one can say is that, given two algebraic closures Ω, Ω' of F, then, assuming the ultrafilter principle, there exists an F-isomorphism $\Omega \to \Omega'$.

Separable closures

Let Ω be a field containing F, and let $\mathcal{E}$ be a set of intermediate fields $F \subset E \subset \Omega$ with the following property:

> (*) for all $E_1, E_2 \in \mathcal{E}$, there exists an $E \in \mathcal{E}$ such that $E_1, E_2 \subset E$.

Then $E(\mathcal{E}) = \bigcup_{E \in \mathcal{E}} E$ is a subfield of Ω (and we call $\bigcup_{E \in \mathcal{E}} E$ a **directed union**), because (*) implies that every finite set of elements of $E(\mathcal{E})$ is contained in a common $E \in \mathcal{E}$, and therefore their product, sum, etc., also lie in $E(\mathcal{E})$.

We apply this remark to the set of subfields E of Ω that are finite and separable over F. As the composite of any two such subfields is again finite and separable over F (3.15), we see that the union L of all such E is a subfield of Ω. Then L is separable over F and every element of Ω separable over F lies in L. Moreover, because a separable extension of a separable extension is separable, Ω is purely inseparable over L.

DEFINITION 6.9 (a) A field Ω is **separably closed** if every nonconstant separable polynomial in $\Omega[X]$ splits in Ω.

(b) A field Ω is a **separable closure** of a subfield F if it is separable and algebraic over F and it is separably closed.

THEOREM 6.10 *(a) Every field has a separable closure.*

(b) Let E be a separable algebraic extension of F, and let Ω be a separable algebraic closure of F. There exists an F-homomorphism $E \to \Omega$, and, if E is also a separable closure of F, then every such homomorphism is an isomorphism.

PROOF. Replace "polynomial" with "separable polynomial" in the proofs of the corresponding theorems for algebraic closures. Alternatively, define Ω to be the separable closure of F in an algebraic closure, and apply the preceding theorems. $\qquad\square$

CHAPTER 7

Infinite Galois Extensions

An algebraic extension Ω, possibly infinite, of a field F is said to be Galois if it is normal and separable. For each finite Galois subextension M/F of Ω, we have a restriction map $\mathrm{Aut}(\Omega/F) \to \mathrm{Gal}(M/F)$, and hence a homomorphism $\mathrm{Aut}(\Omega/F) \to \prod_M \mathrm{Gal}(M/F)$, where the product is over all such subextensions. Clearly every element of Ω lies in some M, and so this homomorphism is injective. When we endow each group $\mathrm{Gal}(M/F)$ with the discrete topology, the product acquires a topology for which it is compact. The image of the homomorphism is closed, and so $\mathrm{Aut}(\Omega/F)$ also acquires a compact topology — we write $\mathrm{Gal}(\Omega/F)$ for $\mathrm{Aut}(\Omega/F)$ endowed with this topology. Now, all of the Galois theory of finite extensions holds for infinite extensions[1] provided "subgroup" is replaced everywhere with "closed subgroup". The reader prepared to accept this, can skip to the examples and exercises.

In this chapter, we make free use of the axiom of choice.[2] We also assume the reader is familiar with infinite topological products, including Tychonoff's theorem.

---

[1] An exception: it need no longer be true that the cardinality of $\mathrm{Gal}(\Omega/F)$ equals the degree $[\Omega:F]$. Certainly, $\mathrm{Gal}(\Omega/F)$ is infinite if and only if $[\Omega:F]$ is infinite, but $\mathrm{Gal}(\Omega/F)$ is always uncountable when infinite whereas $[\Omega:F]$ need not be.

[2] It is necessary to assume some choice axiom in order to have a sensible Galois theory of infinite extensions. For example, it is consistent with Zermelo-Fraenkel set theory that there exist an algebraic closure of $\mathbb{Q}$ with no nontrivial automorphisms. See: Hodges, *Läuchli's algebraic closure of* $\mathbb{Q}$. Math. Proc. Cambridge Philos. Soc. 79 (1976), no. 2, 289–297.

Topological groups

Definition 7.1 A set G together with a group structure and a topology is a *topological group* if the maps

$$(g,h) \mapsto gh \colon G \times G \to G,$$
$$g \mapsto g^{-1} \colon G \to G$$

are both continuous.

Let a be an element of a topological group G. Then $a_L \colon G \xrightarrow{\ g \mapsto ag\ } G$ is continuous because it is the composite of

$$G \xrightarrow{\ g \mapsto (a,g)\ } G \times G \xrightarrow{\ (g,h) \mapsto gh\ } G.$$

In fact, it is a homeomorphism with inverse $(a^{-1})_L$. Similarly $a_R \colon g \mapsto ga$ and $g \mapsto g^{-1}$ are both homeomorphisms. In particular, for any subgroup H of G, the coset aH of H is open or closed according as H is open or closed. Because the complement of H in G is a union of such cosets, this shows that H is closed if it is open, and it is open if it is closed and of finite index.

Recall that a *neighbourhood base* for a point x of a topological space X is a set of neighbourhoods $\mathcal{N}$ such that every open subset U of X containing x contains an N from $\mathcal{N}$.

Proposition 7.2 *Let G be a topological group, and let $\mathcal{N}$ be a neighbourhood base for the identity element e of G. Then*[3]
  (a) *for all $N_1, N_2 \in \mathcal{N}$, there exists an $N' \in \mathcal{N}$ such that $e \in N' \subset N_1 \cap N_2$;*
  (b) *for all $N \in \mathcal{N}$, there exists an $N' \in \mathcal{N}$ such that $N'N' \subset N$;*
  (c) *for all $N \in \mathcal{N}$, there exists an $N' \in \mathcal{N}$ such that $N' \subset N^{-1}$;*
  (d) *for all $N \in \mathcal{N}$ and all $g \in G$, there exists an $N' \in \mathcal{N}$ such that $N' \subset gNg^{-1}$;*
  (e) *for all $g \in G$, $\{gN \mid N \in \mathcal{N}\}$ is a neighbourhood base for g.*

Conversely, if G is a group and $\mathcal{N}$ is a nonempty set of subsets of G satisfying (a,b,c,d), then there is a (unique) topology on G for which (e) holds.

Proof. If $\mathcal{N}$ is a neighbourhood base at e in a topological group G, then (b), (c), and (d) are consequences of the continuity of $(g,h) \mapsto gh$, $g \mapsto g^{-1}$, and

---

[3]For subsets S and S' of G, we let $SS' = \{ss' \mid s \in S,\, s' \in S'\}$ and $S^{-1} = \{s^{-1} \mid s \in S\}$.

$h \mapsto ghg^{-1}$ respectively. Moreover, (a) is a consequence of the definitions and (e) of the fact that g_L is a homeomorphism.

Conversely, let $\mathcal{N}$ be a nonempty collection of subsets of a group G satisfying the conditions (a)–(d). Note that (a) implies that e lies in all the N in $\mathcal{N}$. Define $\mathcal{U}$ to be the collection of subsets U of G such that, for every $g \in U$, there exists an $N \in \mathcal{N}$ with $gN \subset U$. Clearly, the empty set and G are in $\mathcal{U}$, and unions of sets in $\mathcal{U}$ are in $\mathcal{U}$. Let $U_1, U_2 \in \mathcal{U}$, and let $g \in U_1 \cap U_2$; by definition there exist $N_1, N_2 \in \mathcal{N}$ with $gN_1, gN_2 \subset U$; on applying (a) we obtain an $N' \in \mathcal{N}$ such that $gN' \subset U_1 \cap U_2$, which shows that $U_1 \cap U_2 \in \mathcal{U}$. It follows that the elements of $\mathcal{U}$ are the open sets of a topology on G. In fact, it is the unique topology for which (e) holds.

We next use (b) and (d) to show that $(g, g') \mapsto gg'$ is continuous. Note that the sets $g_1 N_1 \times g_2 N_2$ form a neighbourhood base for (g_1, g_2) in $G \times G$. Therefore, given an open $U \subset G$ and a pair (g_1, g_2) such that $g_1 g_2 \in U$, we have to find $N_1, N_2 \in \mathcal{N}$ such that $g_1 N_1 g_2 N_2 \subset U$. As U is open, there exists an $N \in \mathcal{N}$ such that $g_1 g_2 N \subset U$. Apply (b) to obtain an N' such that $N'N' \subset N$; then $g_1 g_2 N'N' \subset U$. But $g_1 g_2 N'N' = g_1 (g_2 N' g_2^{-1}) g_2 N'$, and it remains to apply (d) to obtain an $N_1 \in \mathcal{N}$ such that $N_1 \subset g_2 N' g_2^{-1}$.

Finally, we use (c) and (d) to show that $g \mapsto g^{-1}$ is continuous. Given an open $U \subset G$ and a $g \in G$ such that $g^{-1} \in U$, we have to find an $N \in \mathcal{N}$ such that $gN \subset U^{-1}$. By definition, there exists an $N \in \mathcal{N}$ such that $g^{-1}N \subset U$. Now $N^{-1}g \subset U^{-1}$, and we use (c) to obtain an $N' \in \mathcal{N}$ such that $N'g \subset U^{-1}$, and (d) to obtain an $N'' \in \mathcal{N}$ such that $gN'' \subset g(g^{-1}N'g) \subset U^{-1}$. $\quad\square$

The Krull topology on the Galois group

Recall (3.9) that a finite extension Ω of F is Galois over F if it is normal and separable, i.e., if every irreducible polynomial $f \in F[X]$ having a root in Ω has $\deg f$ distinct roots in Ω. Similarly, we define an algebraic extension Ω of F to be **Galois** over F if it is normal and separable. For example, F^{sep} is a Galois extension of F. Clearly, Ω is Galois over F if and only if it is a directed union of finite Galois extensions.

PROPOSITION 7.3 *If Ω is Galois over F, then it is Galois over every intermediate field M.*

PROOF. Let $f(X)$ be an irreducible polynomial in $M[X]$ having a root a in Ω. The minimal polynomial $g(X)$ of a over F splits into distinct factors of degree one in $\Omega[X]$. As f divides g (in $M[X]$), it also must split into distinct factors of degree one in $\Omega[X]$. $\quad\square$

PROPOSITION 7.4 *Let Ω be a Galois extension of F and let E be a subfield of Ω containing F. Then every F-homomorphism $E \to \Omega$ extends to an F-isomorphism $\Omega \to \Omega$.*

PROOF. The same Zorn's lemma argument as in the proof of Theorem 6.6 shows that every F-homomorphism $E \to \Omega$ extends to an F-homomorphism $\alpha: \Omega \to \Omega$. Let $a \in \Omega$, and let f be its minimal polynomial over F. Then Ω contains exactly $\deg(f)$ roots of f, and so therefore does $\alpha(\Omega)$. Hence $a \in \alpha(\Omega)$, which shows that α is surjective. $\square$

COROLLARY 7.5 *Let $\Omega \supset E \supset F$ be as in the proposition. If E is stable under $\mathrm{Aut}(\Omega/F)$, then E is Galois over F.*

PROOF. Let $f(X)$ be an irreducible polynomial in $F[X]$ having a root a in E. Because Ω is Galois over F, $f(X)$ has $n = \deg(f)$ distinct roots $a_1, \ldots, a_n$ in Ω. There is an F-isomorphism $F[a] \to F[a_i] \subset \Omega$ sending a to a_i (they are both stem fields for f), which extends to an F-isomorphism $\Omega \to \Omega$. As E is stable under $\mathrm{Aut}(\Omega/F)$, this shows that $a_i \in E$. $\square$

Let Ω be a Galois extension of F, and let $G = \mathrm{Aut}(\Omega/F)$. For any finite subset S of Ω, let

$$G(S) = \{\sigma \in G \mid \sigma s = s \text{ for all } s \in S\}.$$

PROPOSITION 7.6 *There is a unique structure of a topological group on G for which the sets $G(S)$ form an open neighbourhood base of 1. For this topology, the sets $G(S)$ with S G-stable form a neighbourhood base of 1 consisting of open normal subgroups.*

PROOF. We show that the collection of sets $G(S)$ satisfies (a,b,c,d) of (7.2). It satisfies (a) because $G(S_1) \cap G(S_2) = G(S_1 \cup S_2)$. It satisfies (b) and (c) because each set $G(S)$ is a group. Let S be a finite subset of Ω. Then $F(S)$ is a finite extension of F, and so there are only finitely many F-homomorphisms $F(S) \to \Omega$. Since $\sigma S = \tau S$ if $\sigma|F(S) = \tau|F(S)$, this shows that $\bar{S} = \bigcup_{\sigma \in G} \sigma S$ is finite. Now $\sigma \bar{S} = \bar{S}$ for all $\sigma \in G$, and it follows that $G(\bar{S})$ is normal in G. Therefore, $\sigma G(\bar{S})\sigma^{-1} = G(\bar{S}) \subset G(S)$, which proves (d). It also proves the second statement. $\square$

The topology on $\mathrm{Aut}(\Omega/F)$ defined in the proposition is called the **Krull topology**. We write $\mathrm{Gal}(\Omega/F)$ for $\mathrm{Aut}(\Omega/F)$ endowed with the Krull topology, and call it the **Galois group** of Ω/F. The Galois group of F^{sep} over F is called the **absolute Galois group**[4] of F.

If S is a finite set stable under G, then $F(S)$ is a finite extension of F stable under G, and hence Galois over F (7.5). Therefore,

$$\{\text{Gal}(\Omega/E) \mid E \text{ finite and Galois over } F\}$$

is a neighbourhood base of 1 consisting of open normal subgroups.

PROPOSITION 7.7 *Let Ω be Galois over F. For every intermediate field E finite and Galois over F, the map*

$$\sigma \mapsto \sigma|E : \text{Gal}(\Omega/F) \to \text{Gal}(E/F)$$

is a continuous surjection (discrete topology on $\text{Gal}(E/F)$).

PROOF. Let $\sigma \in \text{Gal}(E/F)$, and regard it as an F-homomorphism $E \to \Omega$. Then σ extends to an F-isomorphism $\Omega \to \Omega$ (see 7.4), which shows that the map is surjective. For every finite set S of generators of E over F, $\text{Gal}(\Omega/E) = G(S)$, which shows that the inverse image of $1_{\text{Gal}(E/F)}$ is open in G. By homogeneity, the same is true for every element of $\text{Gal}(E/F)$. $\square$

PROPOSITION 7.8 *The Galois group G of a Galois extension Ω/F is compact and totally disconnected.*[5]

PROOF. We first show that G is Hausdorff. If $\sigma \neq \tau$, then $\sigma^{-1}\tau \neq 1_G$, and so it moves some element of Ω, i.e., there exists an $a \in \Omega$ such that $\sigma(a) \neq \tau(a)$. For any S containing a, $\sigma G(S)$ and $\tau G(S)$ are disjoint because their elements act differently on a. Hence they are disjoint open subsets of G containing σ and τ respectively.

We next show that G is compact. As we noted above, if S is a finite set stable under G, then $G(S)$ is a normal subgroup of G, and it has finite index because it is the kernel of

$$G \to \text{Sym}(S).$$

---

[4]But note that the absolute Galois group of F is only defined up to an inner automorphism: let F' be a second separable algebraic closure of F; the choice of an isomorphism $F' \to F^{\text{sep}}$ determines an isomorphism $\text{Gal}(F'/F) \to \text{Gal}(F^{\text{sep}}/F)$; a second isomorphism $F' \to F^{\text{sep}}$ will differ from the first by an element σ of $\text{Gal}(F^{\text{sep}}/F)$, and the isomorphism $\text{Gal}(F'/F) \to \text{Gal}(F^{\text{sep}}/F)$ it defines differs from the first by $\text{inn}(\sigma)$.

[5]A topological space is **totally disconnected** if its connected components are the one-point sets.

Since every finite set is contained in a stable finite set,[6] the argument in the last paragraph shows that the map

$$G \to \prod_{S \text{ finite stable under } G} G/G(S)$$

is injective. When we endow $\prod_S G/G(S)$ with the product topology, the induced topology on G is that for which the $G(S)$ form an open neighbourhood base of e, i.e., it is the Krull topology. According to the Tychonoff theorem, $\prod_S G/G(S)$ is compact, and so it remains to show that G is closed in the product. For each $S_1 \subset S_2$, there are two continuous maps $\prod_S G/G(S) \to G/G(S_1)$, namely, the projection onto $G/G(S_1)$ and the projection onto $G/G(S_2)$ followed by the quotient map $G/G(S_2) \to G/G(S_1)$. Let $E(S_1, S_2)$ be the closed subset of $\prod G/G(S)$ on which the two maps agree. Then $\bigcap_{S_1 \subset S_2} E(S_1, S_2)$ is closed, and equals the image of G.

Finally, for each finite set S stable under G, $G(S)$ is a subgroup that is open and hence closed. Since $\bigcap G(S) = \{1_G\}$, this shows that the connected component of G containing 1_G is just $\{1_G\}$. By homogeneity, a similar statement is true for every element of G. $\qquad\square$

PROPOSITION 7.9 *For every Galois extension Ω/F, $\Omega^{\mathrm{Gal}(\Omega/F)} = F$.*

PROOF. Every element of $\Omega \smallsetminus F$ lies in a finite Galois extension of F, and so this follows from the surjectivity in Proposition 7.7. $\qquad\square$

The next result is an infinite version of Emil Artin's fundamental result 3.11.

PROPOSITION 7.10 *Let G be a group of automorphisms of a field E, and let $F = E^G$. If G is compact and the stabilizer of each element of E is open in G, then E is a Galois extension of F with Galois group G.*

PROOF. Let $x_1, \ldots, x_n$ be a finite set of elements of E, and let H_i be the open subgroup of G fixing x_i. Because G is compact, the orbit Gx_i of x_i is finite, and the subgroups of G fixing its elements are the conjugates of H. Let N be the intersection of all the conjugates of the H_i. It is an open normal subgroup of G, and its fixed field M is that generated over F by the elements of the orbits of the x_i. Thus, G/N is a (finite) group of automorphisms of M with fixed field F. According to 3.11, M is a finite Galois extension of F with Galois group G/N.

---

[6]Each element of Ω is algebraic over F, and its orbit is the set of its conjugates (roots of its minimal polynomial over F), which is finite.

As E is a directed union of such fields M, it is a Galois extension of F. Thus, $\mathrm{Gal}(E/F)$ is defined and, by assumption, G maps continuously and injectively into it. As G is compact, its image is closed, and it is also dense because it maps onto all the group $\mathrm{Gal}(M/F)$. Thus, $G \to \mathrm{Gal}(E/F)$ is an isomorphism. $\qquad\square$

ASIDE 7.11 Not all compact totally disconnected group arise as the *absolute* Galois group of a field. In fact, absolute Galois groups of fields of characteristic zero, if finite, must have order 1 or 2. More precisely, there is the following theorem of Artin and Schreier (1927): let F be a field, not algebraically closed, but of finite index in its algebraic closure; then F is real-closed and $E = F[\sqrt{-1}]$ (Jacobson, *Lectures in Abstract Algebra*, 1964, Vol. III, Chap. VI, Theorem 17).

The fundamental theorem of infinite Galois theory

PROPOSITION 7.12 *Let Ω be Galois over F, with Galois group G.*

(a) *Let M be a subfield of Ω containing F. Then Ω is Galois over M, the Galois group $\mathrm{Gal}(\Omega/M)$ is closed in G, and $\Omega^{\mathrm{Gal}(\Omega/M)} = M$.*

(b) *For every subgroup H of G, $\mathrm{Gal}(\Omega/\Omega^H)$ is the closure of H.*

PROOF. (a) The first assertion was proved in (7.3). For each finite subset $S \subset M$, $G(S)$ is an open subgroup of G, and hence it is closed. But $\mathrm{Gal}(\Omega/M) = \bigcap_{S \subset M} G(S)$, and so it also is closed. The final statement now follows from (7.9).

(b) Since $\mathrm{Gal}(\Omega/\Omega^H)$ contains H and is closed, it certainly contains the closure $\bar{H}$ of H. On the other hand, let $\sigma \in G \smallsetminus \bar{H}$; we have to show that σ moves some element of Ω^H. Because σ is not in the closure of H,

$$\sigma\,\mathrm{Gal}(\Omega/E) \cap H = \emptyset$$

for some finite Galois extension E of F in Ω (because the sets $\mathrm{Gal}(\Omega/E)$ form a neighbourhood base of 1; see above). Let ϕ denote the surjective map $\mathrm{Gal}(\Omega/F) \to \mathrm{Gal}(E/F)$. Then $\sigma|E \notin \phi H$, and so σ moves some element of $E^{\phi H} \subset \Omega^H$ (apply 3.11). $\qquad\square$

THEOREM 7.13 *Let Ω be a Galois extension of F with Galois group G. The maps*

$$H \mapsto \Omega^H, \quad M \mapsto \mathrm{Gal}(\Omega/M)$$

are inverse bijections between the set of closed subgroups of G and the set of intermediate fields between Ω and F:

$$\{\text{closed subgroups of } G\} \overset{1:1}{\longleftrightarrow} \{\text{intermediate fields } F \subset M \subset \Omega\}.$$

Moreover,

(a) $H_1 \supset H_2 \iff \Omega^{H_1} \subset \Omega^{H_2}$ *(the correspondence is order reversing)*;

(b) *a closed subgroup H of G is open if and only if Ω^H has finite degree over F, in which case $(G:H) = [\Omega^H : F]$;*

(c) $\sigma H \sigma^{-1} \leftrightarrow \sigma M$, *i.e.,*

$$\Omega^{\sigma H \sigma^{-1}} = \sigma(\Omega^H);$$

$$\mathrm{Gal}(\Omega/\sigma M) = \sigma\, \mathrm{Gal}(\Omega/M)\sigma^{-1};$$

(d) *a closed subgroup H of G is normal if and only if Ω^H is Galois over F, in which case*

$$\mathrm{Gal}(\Omega^H/F) \simeq G/H.$$

PROOF. For the first statement, we have to show that $H \mapsto \Omega^H$ and $M \mapsto \mathrm{Gal}(\Omega/M)$ are inverse maps.

Let H be a closed subgroup of G. Then Ω is Galois over Ω^H and $\mathrm{Gal}(\Omega/\Omega^H) = H$ (see 7.12).

Let M be an intermediate field. Then $\mathrm{Gal}(\Omega/M)$ is a closed subgroup of G and $\Omega^{\mathrm{Gal}(\Omega/M)} = M$ (see 7.12).

(a) We have the obvious implications:

$$H_1 \supset H_2 \implies \Omega^{H_1} \subset \Omega^{H_2} \implies \mathrm{Gal}(\Omega/\Omega^{H_1}) \supset \mathrm{Gal}(\Omega/\Omega^{H_2}).$$

But $\mathrm{Gal}(\Omega/\Omega^{H_i}) = H_i$ (see 7.12).

(b) As we noted earlier, a closed subgroup of finite index in a topological group is always open. Because G is compact, conversely an open subgroup of G is always of finite index. Let H be such a subgroup. The map $\sigma \mapsto \sigma|\Omega^H$ defines a bijection

$$G/H \to \mathrm{Hom}_F(\Omega^H, \Omega)$$

(apply 7.4) from which the statement follows.

(c) For $\tau \in G$ and $\alpha \in \Omega$, $\tau\alpha = \alpha \iff \sigma\tau\sigma^{-1}(\sigma\alpha) = \sigma\alpha$. Therefore, $\mathrm{Gal}(\Omega/\sigma M) = \sigma\, \mathrm{Gal}(\Omega/M)\sigma^{-1}$, and so $\sigma\, \mathrm{Gal}(\Omega/M)\sigma^{-1} \leftrightarrow \sigma M$.

(d) Let $H \leftrightarrow M$. It follows from (c) that H is normal if and only if M is stable under the action of G. But M is stable under the action of G if and only it is a union of finite extensions of F stable under G, i.e., of finite Galois extensions of G. We have already observed that an extension is Galois if and only if it is a union of finite Galois extensions. $\square$

REMARK 7.14 As in the finite case (3.18), we can deduce the following statements.

(a) Let $(M_i)_{i \in I}$ be a (possibly infinite) family of intermediate fields, and let $H_i \leftrightarrow M_i$. Let M be the smallest field containing all the M_i (the composite of the M_i); then because $\bigcap_{i \in I} H_i$ is the largest (closed) subgroup contained in all the H_i,

$$\mathrm{Gal}(\Omega/M) = \bigcap_{i \in I} H_i.$$

(b) Let $M \leftrightarrow H$. The largest (closed) normal subgroup contained in H is $N = \bigcap_\sigma \sigma H \sigma^{-1}$ (cf. GT, 4.10), and so Ω^N, which is the composite of the fields σM, is the smallest normal extension of F containing M.

PROPOSITION 7.15 *Let E and L be field extensions of F contained in some common field. If E/F is Galois, then EL/L and $E/E \cap L$ are Galois, and the map*

$$\sigma \mapsto \sigma|E : \mathrm{Gal}(EL/L) \to \mathrm{Gal}(E/E \cap L)$$

is an isomorphism of topological groups.

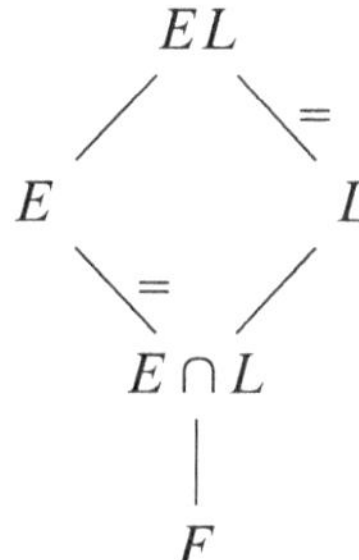

PROOF. We first prove that the map is continuous. Let $G_1 = \mathrm{Gal}(EL/L)$ and let $G_2 = \mathrm{Gal}(E/E \cap L)$. For any finite set S of elements of E, the inverse image of $G_2(S)$ in G_1 is $G_1(S)$.

We next show that the map is an isomorphism of groups (neglecting the topology). As in the finite case, it is an injective homomorphism (3.19). Let H be the image of the map. Then the fixed field of H is $E \cap L$, which implies that H is dense in $\mathrm{Gal}(E/E \cap L)$. But H is closed because it is the continuous image of a compact space in a Hausdorff space, and so $H = \mathrm{Gal}(E/E \cap L)$.

Finally, we prove that it is open. An open subgroup of $\mathrm{Gal}(EL/L)$ is closed (hence compact) of finite index; therefore its image in $\mathrm{Gal}(E/E \cap L)$ is compact (hence closed) of finite index, and hence open. $\qquad\square$

COROLLARY 7.16 *Let Ω be an algebraically closed field containing F, and let E and L be as in the proposition. If $\rho: E \to \Omega$ and $\sigma: L \to \Omega$ are F-homomorphisms such that $\rho|E \cap L = \sigma|E \cap L$, then there exists an F-homomorphism $\tau: EL \to \Omega$ such that $\tau|E = \rho$ and $\tau|L = \sigma$.*

PROOF. According to (7.4), σ extends to an F-homomorphism $s\colon EL \to \Omega$. As $s|E \cap L = \rho|E \cap L$, we can write $s|E = \rho \circ \varepsilon$ for some $\varepsilon \in \operatorname{Gal}(E/E \cap L)$. According to the proposition, there exists a unique $e \in \operatorname{Gal}(EL/L)$ such that $e|E = \varepsilon$. Define $\tau = s \circ e^{-1}$. $\qquad\square$

EXAMPLE 7.17 Let Ω be an algebraic closure of the finite field $\mathbb{F}_p$. Then $G = \operatorname{Gal}(\Omega/\mathbb{F}_p)$ contains a canonical Frobenius element, $\sigma = (a \mapsto a^p)$, and it is generated by it as a topological group, i.e., G is the closure of $\langle \sigma \rangle$. We now determine the structure of G.

Endow $\mathbb{Z}$ with the topology for which the groups $n\mathbb{Z}$, $n \geq 1$, form a fundamental system of neighbourhoods of 0. Thus two integers are close if their difference is divisible by a large integer.

As for any topological group, we can complete $\mathbb{Z}$ for this topology. A Cauchy sequence in $\mathbb{Z}$ is a sequence $(a_i)_{i \geq 1}$, $a_i \in \mathbb{Z}$, satisfying the following condition: for all $n \geq 1$, there exists an N such that $a_i \equiv a_j \mod n$ for $i, j > N$. Call a Cauchy sequence in $\mathbb{Z}$ trivial if $a_i \to 0$ as $i \to \infty$, i.e., if for all $n \geq 1$, there exists an N such that $a_i \equiv 0 \mod n$ for all $i > N$. The Cauchy sequences form a commutative group, and the trivial Cauchy sequences form a subgroup. We define $\hat{\mathbb{Z}}$ to be the quotient of the first group by the second. It has a ring structure, and the map sending $m \in \mathbb{Z}$ to the constant sequence $m, m, m, \ldots$ identifies $\mathbb{Z}$ with a subgroup of $\hat{\mathbb{Z}}$.

Let $\alpha \in \hat{\mathbb{Z}}$ be represented by the Cauchy sequence (a_i). The restriction of the Frobenius element σ to $\mathbb{F}_{p^n}$ has order n. Therefore $(\sigma|\mathbb{F}_{p^n})^{a_i}$ is independent of i provided it is sufficiently large, and we can define $\sigma^\alpha \in \operatorname{Gal}(\Omega/\mathbb{F}_p)$ to be such that, for each n, $\sigma^\alpha|\mathbb{F}_{p^n} = (\sigma|\mathbb{F}_{p^n})^{a_i}$ for all i sufficiently large (depending on n). The map $\alpha \mapsto \sigma^\alpha \colon \hat{\mathbb{Z}} \to \operatorname{Gal}(\Omega/\mathbb{F}_p)$ is an isomorphism.

The group $\hat{\mathbb{Z}}$ is uncountable. To most analysts, it is a little weird—its connected components are one-point sets. To number theorists it will seem quite natural — the Chinese remainder theorem implies that it is isomorphic to $\prod_{p \text{ prime}} \mathbb{Z}_p$ where $\mathbb{Z}_p$ is the ring of p-adic integers.

EXAMPLE 7.18 Let $\mathbb{Q}^{\mathrm{al}}$ be the algebraic closure of $\mathbb{Q}$ in $\mathbb{C}$. Then $\operatorname{Gal}(\mathbb{Q}^{\mathrm{al}}/\mathbb{Q})$ is one of the most basic, and intractable, objects in mathematics. It is expected that *every* finite group occurs as a quotient of it. This is known, for example, for S_n and for every sporadic simple group except possibly M_{23}. See (5.41) and mo80359.

On the other hand, we do understand $\operatorname{Gal}(F^{\mathrm{ab}}/F)$, where $F \subset \mathbb{Q}^{\mathrm{al}}$ is a finite extension of $\mathbb{Q}$ and F^{ab} is the union of all finite abelian extensions of F contained in $\mathbb{Q}^{\mathrm{al}}$. For example, $\operatorname{Gal}(\mathbb{Q}^{\mathrm{ab}}/\mathbb{Q}) \simeq \hat{\mathbb{Z}}^\times$. This is abelian class field theory — see my notes *Class Field Theory*.

ASIDE 7.19 A *simple Galois correspondence* is a system consisting of two partially ordered sets P and Q and order reversing maps $f: P \to Q$ and $g: Q \to P$ such that $gf(p) \geq p$ for all $p \in P$ and $fg(q) \geq q$ for all $q \in Q$. Then $fgf = f$, because $fg(fp) \geq fp$ and $gf(p) \geq p$ implies $f(gfp) \leq f(p)$ for all $p \in P$. Similarly, $gfg = g$, and it follows that f and g define a one-to-one correspondence between the sets $g(Q)$ and $f(P)$.

From a Galois extension Ω of F we get a simple Galois correspondence by taking P to be the set of subgroups of $\mathrm{Gal}(\Omega/F)$ and Q to be the set of subsets of Ω, and by setting $f(H) = \Omega^H$ and $g(S) = G(S)$. Thus, to prove the one-to-one correspondence in the fundamental theorem, it suffices to identify the closed subgroups as exactly those in the image of g and the intermediate fields as exactly those in the image of f. This is accomplished by (7.12).

Galois groups as inverse limits

DEFINITION 7.20 A partial order $\leq$ on a set I is *directed*, and the pair $(I, \leq)$ is a *directed set*, if for all $i, j \in I$ there exists a $k \in I$ such that $i, j \leq k$.

DEFINITION 7.21 Let $(I, \leq)$ be a directed set, and let C be a category (for example, the category of groups and homomorphisms, or the category of topological groups and continuous homomorphisms).

(a) An *inverse system* in C indexed by $(I, \leq)$ is a family $(A_i)_{i \in I}$ of objects of C together with a family $(p_i^j: A_j \to A_i)_{i \leq j}$ of morphisms such that $p_i^i = \mathrm{id}_{A_i}$ and $p_i^j \circ p_j^k = p_i^k$ all $i \leq j \leq k$.

(b) An object A of C together with a family $(p_j: A \to A_j)_{j \in I}$ of morphisms satisfying $p_i^j \circ p_j = p_i$ all $i \leq j$ is an *inverse limit* of the system in (a) if it has the following universal property: for any other object B and family $(q_j: B \to A_j)$ of morphisms such $p_i^j \circ q_j = q_i$ all $i \leq j$, there exists a unique morphism $r: B \to A$ such that $p_j \circ r = q_j$ for j,

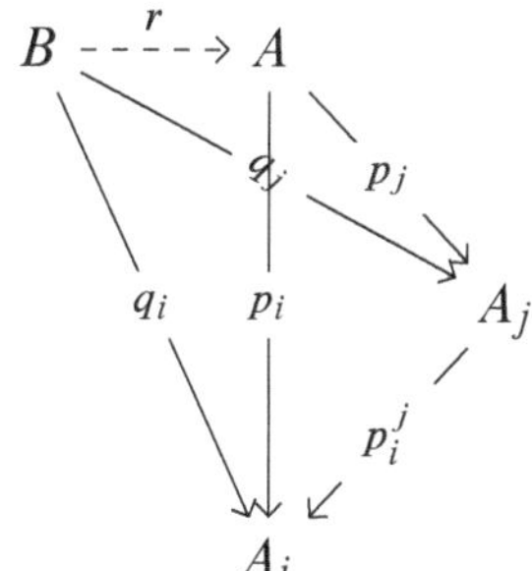

Clearly, the inverse limit (if it exists), is uniquely determined by this condition up to a unique isomorphism. We denote it by $\varprojlim(A_i, p_i^j)$, or just $\varprojlim A_i$.

EXAMPLE 7.22　Let $(G_i, p_i^j\colon G_j \to G_i)$ be an inverse system of groups. Let

$$G = \{(g_i) \in \prod G_i \mid p_i^j(g_j) = g_i \text{ all } i \leq j\},$$

and let $p_i\colon G \to G_i$ be the projection map. Then $p_i^j \circ p_j = p_i$ is just the equation $p_i^j(g_j) = g_i$. Let (H, q_i) be a second family such that $p_i^j \circ q_j = q_i$. The image of the homomorphism

$$h \mapsto (q_i(h))\colon H \to \prod G_i$$

is contained in G, and this is the unique homomorphism $H \to G$ carrying q_i to p_i. Hence $(G, p_i) = \varprojlim(G_i, p_i^j)$.

EXAMPLE 7.23　Let $(G_i, p_i^j\colon G_j \to G_i)$ be an inverse system of topological groups and continuous homomorphisms. When endowed with the product topology, $\prod G_i$ becomes a topological group, and G becomes a topological subgroup with the subspace topology,

$$G = \{(g_i) \in \prod G_i \mid p_i^j(g_j) = g_i \text{ all } i \leq j\}.$$

The projection maps p_i are continuous. Let H be (H, q_i) be a second family such that $p_i^j \circ q_j = q_i$. The homomorphism

$$h \mapsto (q_i(h))\colon H \to \prod G_i$$

is continuous because its composites with projection maps are continuous (universal property of the product). Therefore $H \to G$ is continuous, and this shows that $(G, p_i) = \varprojlim(G_i, p_i^j)$.

An inverse system of finite groups can be regarded as an inverse system of topological groups by giving each finite group the discrete topology.

DEFINITION 7.24　A ***profinite group*** is topological group that is an inverse limit of finite groups (each equipped with the discrete topology).

Inverse limits are also called projective limits. Thus "profinite group" is short for "projective limit of finite groups".

PROPOSITION 7.25 *A topological group is profinite if and only if it is compact and totally disconnected.*

PROOF. Let $(G_i, p_i^j : G_j \to G_i)$ be an inverse system of finite groups, and let $G = \varprojlim G_i$. Thus,

$$G = \{(g_i) \in \prod G_i \mid p_i^j(g_j) = g_i \text{ all } i \le j\}.$$

If $(x_i) \notin G$, say $p_{i_0}^{j_0}(x_{j_0}) \ne x_{i_0}$, then

$$G \cap \{(g_j) \mid g_{j_0} = x_{j_0}, \quad g_{i_0} = x_{i_0}\} = \emptyset.$$

As the second set is an open neighbourhood of (x_i), this shows that G is closed in $\prod G_i$. By Tychonoff's theorem, $\prod G_i$ is compact, and so G is also compact. The map $p_i : G \to G_i$ is continuous, and its kernel U_i is an open subgroup of finite index in G (hence also closed). As $\bigcap U_i = \{e\}$, the connected component of G containing e is just $\{e\}$. By homogeneity, the same is true for every point of G: the connected components of G are the one-point sets — G is totally disconnected.

Conversely, let G be compact and totally disconnected. In a locally compact group, the connected component containing the identity is the intersection of the open subgroups (Bourbaki, *Topologie Générale*, Chap. III, §4.6). Therefore, $\bigcap U = \{e\}$ in G (intersection over the open subgroups). As G is compact, each U has finite index; therefore its conjugates are finite in number, and their intersection is a normal open subgroup of G. Hence $\bigcap V = \{1\}$ in G (intersection over the open normal subgroups). The canonical map $G \to \varprojlim G/V$ is injective, continuous, with dense image. As G is compact, it is an isomorphism. $\qquad\square$

EXAMPLE 7.26 Let Ω be a Galois extension of F. The composite of two finite Galois extensions of in Ω is again a finite Galois extension (3.21), and so the finite Galois subextensions of Ω form a directed set I. For each E in I we have a finite group $\mathrm{Gal}(E/F)$, and for each $E \subset E'$ we have a restriction homomorphism $p_E^{E'} : \mathrm{Gal}(E'/F) \to \mathrm{Gal}(E/F)$. In this way, we get an inverse system of finite groups $(\mathrm{Gal}(E/F), p_E^{E'})$ indexed by I.

For each E, there is a restriction homomorphism $p_E : \mathrm{Gal}(\Omega/F) \to \mathrm{Gal}(E/F)$ and, because of the universal property of inverse limits, these maps define a homomorphism

$$\mathrm{Gal}(\Omega/F) \to \varprojlim \mathrm{Gal}(E/F).$$

This map is an isomorphism of topological groups. This is a restatement of what we showed in the proof of (7.8).

PROPOSITION 7.27 (TATE) *Every profinite group G is the Galois group of some Galois extension of fields.*

PROOF. Let S be the disjoint union of the sets G/H for H an open subgroup of G. Then G acts faithfully on S and the stabilizer of each element of S is open in G. Let k be a field, let $k[S]$ be the polynomial ring over k in the elements of S, and let $E = k(S)$ be the field of fractions of $k[S]$. Then G acts faithfully on E through its action on S and the stabilizer of each element of E is open in G. According to Proposition 7.10, E is Galois over $F \overset{\text{def}}{=} E^G$ with Galois group G. $\qquad\square$

Nonopen subgroups of finite index

We apply Zorn's lemma to construct a nonopen subgroup of finite index in $\mathrm{Gal}(\mathbb{Q}^{\mathrm{al}}/\mathbb{Q})$.

LEMMA 7.28 *Let V be an infinite-dimensional vector space. For all $n \geq 1$, there exists a subspace V_n of V such that V/V_n has dimension n.*

PROOF. A Zorn's lemma argument shows that V contains maximal linearly independent subsets, and then the usual argument shows that such a subset spans V, i.e., is a basis. Choose a basis, and take V_n to be the subspace spanned by the set obtained by omitting n elements from the basis. $\qquad\square$

PROPOSITION 7.29 *The group $\mathrm{Gal}(\mathbb{Q}^{\mathrm{al}}/\mathbb{Q})$ has nonopen normal subgroups of index 2^n for all $n > 1$.*

PROOF. Let E be the subfield of $\mathbb{C}$ generated over $\mathbb{C}$ by $\sqrt{-1}$ and the square roots of the prime numbers — it is Galois over $\mathbb{Q}$. For each p,

$$\mathrm{Gal}(\mathbb{Q}[\sqrt{-1}, \sqrt{2}, \ldots, \sqrt{p}]/\mathbb{Q})$$

is a product of copies of $\mathbb{Z}/2\mathbb{Z}$ indexed by the set $\{\text{primes} \leq p\} \cup \{\infty\}$ (see 5.31b). As

$$\mathrm{Gal}(E/\mathbb{Q}) = \varprojlim \mathrm{Gal}(\mathbb{Q}[\sqrt{-1}, \sqrt{2}, \ldots, \sqrt{p}]/\mathbb{Q}),$$

it is a direct product of copies of $\mathbb{Z}/2\mathbb{Z}$ indexed by the primes l of $\mathbb{Q}$ (including $l = \infty$) endowed with the product topology. Let $G = \mathrm{Gal}(E/\mathbb{Q})$, and let

$$H = \{(a_l) \in G \mid a_l = 0 \text{ for all but finitely many } l\}.$$

This is a subgroup of G (in fact, it is a direct *sum* of copies of $\mathbb{Z}/2\mathbb{Z}$ indexed by the primes of $\mathbb{Q}$), and it is dense in G: let $(a_l) \in G$; then the sequence

$$(a_\infty,0,0,0,\ldots), (a_\infty,a_2,0,0,\ldots), (a_\infty,a_2,a_3,0,\ldots),\ldots$$

in H converges to (a_l). We can regard G/H as vector space over $\mathbb{F}_2$ and apply the lemma to obtain subgroups G_n of index 2^n in G containing H. If G_n is open in G, then it is closed, which contradicts the fact that H is dense. Therefore, G_n is not open, and its inverse image in $\mathrm{Gal}(\mathbb{Q}^{\mathrm{al}}/\mathbb{Q})$ is the desired subgroup (if it were open, it would be closed of finite index, and so would its image G_n). $\qquad\qquad\qquad\qquad\qquad\qquad\qquad\qquad\qquad\qquad\qquad\qquad\quad\square$

REMARK 7.30 Let $G = \mathrm{Gal}(\mathbb{Q}^{\mathrm{al}}/\mathbb{Q})$. We showed in the above proof that there is a closed normal subgroup $N = \mathrm{Gal}(\mathbb{Q}^{\mathrm{al}}/E)$ of G such that G/N is an uncountable vector space over $\mathbb{F}_2$. Let $(G/N)^\vee$ be the dual of this vector space (also uncountable). Every nonzero $f \in (G/N)^\vee$ defines a surjective map $G \to \mathbb{F}_2$ whose kernel is a subgroup of index 2 in G. These subgroups are distinct, and so G has uncountably many subgroups of index 2. Only countably many of them are open because $\mathbb{Q}$ has only countably many quadratic extensions in a fixed algebraic closure.

ASIDE 7.31 Zorn's lemma is needed for 7.29 — it is consistent with ZF+DC (dependent choice) that every homomorphism from a second countable profinite group to a finite group be continuous (see mo106216).

ASIDE 7.32 Let G be a profinite group that is finitely generated as a topological group. It is a difficult theorem, only proved this century, that every subgroup of finite index in G is open (Nikolov, Segal, *On finitely generated profinite groups. I.*, Ann. of Math. (2) 165 (2007), no. 1, 171–238.)

Exercises

7-1 Let p be a prime number, and let Ω be the subfield of $\mathbb{C}$ generated over $\mathbb{Q}$ by all p^mth roots of 1 for $m \in \mathbb{N}$. Show that Ω is Galois over $\mathbb{Q}$ with Galois group $\mathbb{Z}_p^\times = \varprojlim(\mathbb{Z}/p^m\mathbb{Z})^\times$. Hint: Use that Ω is the union of a tower of subfields

$$\mathbb{Q} \subset \mathbb{Q}[\zeta_p] \subset \cdots \subset \mathbb{Q}[\zeta_{p^m}] \subset \mathbb{Q}[\zeta_{p^{m+1}}] \subset \cdots.$$

For p odd, show that $\mathrm{Gal}(\mathbb{Q}(\zeta_{p^\infty})/\mathbb{Q}(\zeta_p)) \simeq \mathbb{Z}_p$. Hint: Let $a \in \mathbb{Z}_p$ correspond to

$$\zeta_{p^k} \mapsto \zeta_{p^k}^{(1+p)^{a \bmod p^{k-1}}}.$$

7-2 Let $\mathbb{F}$ be an algebraic closure of $\mathbb{F}_p$, and let $\mathbb{F}_{p^m}$ be the subfield of $\mathbb{F}$ with p^m elements. Show that

$$\varprojlim_{m \geq 1} \mathrm{Gal}(\mathbb{F}_{p^m}/\mathbb{F}_p) \simeq \varprojlim_{m \geq 1} \mathbb{Z}/m\mathbb{Z}$$

and deduce that $\mathrm{Gal}(\mathbb{F}/\mathbb{F}_p) \simeq \hat{\mathbb{Z}}$.

7-3 For a profinite group G, define G^{ab} to be the quotient of G by the closure of its commutator subgroup. Is $G^{\mathrm{ab}} = \varprojlim_i G_i^{\mathrm{ab}}$ where the G_i range over the finite quotients of G.

CHAPTER 8

The Galois theory of étale algebras

For Grothendieck, the classification of field extensions by Galois groups, and the classification of covering spaces by fundamental groups, are two aspects of the same theory. In this chapter, we re-interprete classical Galois theory from Grothendieck's point of view. We assume the reader is familiar with the language of category theory (Wikipedia: category theory; equivalence of categories).

Throughout, F is a field, all rings and F-algebras are commutative, and unadorned tensor products are over F. Recall that an F-algebra A is finite if it is finite-dimensional as an F-vector space — the dimension is called the degree $[A:F]$ of A.

Review of commutative algebra

We'll need some results from commutative algebra. The first is a special case of the Chinese Remainder Theorem.

THEOREM 8.1 *Let $M_1,\dots M_n$ be maximal ideals in a ring A. Then the map*

$$a \mapsto (\dots, a + M_j, \dots): A \to A/M_1 \times \cdots \times A/M_n \qquad (13)$$

is surjective with kernel $\bigcap_i M_i$.

PROOF. Fix a j and, for $i \neq j$, let $a_{ij} \in M_i \smallsetminus M_j$. After scaling, we may suppose that $a_{ij} \equiv 1 \bmod M_j$. Let $b_j = \prod_{i \neq j} a_{ij}$. Then b_j maps to $(0,\ldots,0,1,0,\ldots,0)$ in $\prod_i A/M_i$, and so every element of $\prod_i A/M_i$ is the image of an element $\sum_j a_j b_j$ of A. We have shown that the map is surjective, and its kernel is obviously $\bigcap_i M_i$. $\qquad\square$

The **radical** of an ideal I in a ring A is the set of $f \in A$ such that $f^n \in I$ for some $n \in \mathbb{N}$. It is again an ideal, and it is equal to its own radical. The **nilradical** N of A is the radical of the ideal (0). It consists of the nilpotents in A. If $N = 0$, then A is said to be **reduced**.

THEOREM 8.2 *Let A be a finitely generated F-algebra, and let I be an ideal in A. The radical of I is equal to the intersection of the maximal ideals containing it,*

$$\mathrm{rad}(I) = \bigcap \{M \mid M \supset I,\, M \text{ maximal}\}.$$

In particular, A is reduced if and only if $\bigcap \{M \mid M \text{ maximal}\} = 0$.

We prove this for finite F-algebras, which is the only case we'll need. Let Ω be an algebraic closure of F. For an ideal I of $k[X_1,\ldots,X_n]$, we let $Z(I)$ denote the zero set,

$$Z(I) = \{(a_1,\ldots,a_n) \in \Omega^n \mid f(a_1,\ldots,a_n) = 0 \text{ for all } f \in I\}.$$

LEMMA 8.3 *Let I be a proper ideal of $F[X_1,\ldots,X_n]$ such that the quotient $F[X_1,\ldots,X_n]/I$ is a finite F-algebra. Then $Z(I) \neq \emptyset$.*

PROOF. Let M be a maximal ideal containing I (any proper ideal of largest dimension will do). Then $F[X_1,\ldots,X_n]/M$ is field, finite over F, and so it admits a homomorphism into Ω. Let a_i denote the image of X_i in Ω. Then $(a_1,\ldots,a_n) \in Z(I)$. $\qquad\square$

LEMMA 8.4 *Let I be a proper ideal of $F[X_1,\ldots,X_n]$ such that the quotient $F[X_1,\ldots,X_n]/I$ is a finite F-algebra. If $h \in F[X_1,\ldots,X_n]$ is zero on $Z(I)$, then some power of h lies in I.*

PROOF. We may suppose $h \neq 0$. Then 8.4 can be deduced from 8.3 by using Rabinowitsch's trick (CA 13.10). $\qquad\square$

We now prove Theorem 8.2 in the case that A is a finite F-algebra. Because of the correspondence between ideals in a ring and in a quotient of the ring, we may suppose that I is an ideal in $F[X_1,\ldots,X_n]$. The inclusion

$$\mathrm{rad}(I) \subset \bigcap \{M \mid M \supset I,\, M \text{ maximal}\}$$

holds in any ring because maximal ideals are radical and $\mathrm{rad}(I)$ is the smallest radical ideal containing I. For the reverse inclusion, let h lie in all maximal ideals containing I, and let $(a_1,\ldots,a_n) \in Z(I)$. The image of the evaluation map

$$f \mapsto f(a_1,\ldots,a_n) \colon F[X_1,\ldots,X_n] \to \Omega$$

is a subring of Ω, finite over F, and so a field (1.23). Therefore, the kernel of the map is a maximal ideal, which contains I, and hence also h. This shows that $h(a_1,\ldots,a_n) = 0$, and we conclude from 8.4 that $h \in \mathrm{rad}(I)$.

Étale algebras over a field

Let $F^n = F \times \cdots \times F$ (n-copies) regarded as an F-algebra by the diagonal map.

DEFINITION 8.5 An F-algebra A is **diagonalizable** if it is isomorphic to F^n for some n, and it is **étale** if $L \otimes A$ is diagonalizable for some field L containing F.[1]

Let A be a finite F-algebra. For any finite set S of maximal ideals in A, the Chinese remainder theorem (8.1) shows that the map $A \to \prod_{M \in S} A/M$ is surjective with kernel $\bigcap_{M \in S} M$. In particular, $|S| \le [A \colon F]$, and so A has only finitely many maximal ideals. If S is the set of all maximal ideals in A, then $\bigcap_{M \in S} M$ is the nilradical N of A (8.2), and so A/N is a finite product of fields.

PROPOSITION 8.6 *The following conditions on a finite F-algebra A are equivalent:*

(a) *A is étale;*

(b) *$L \otimes A$ is reduced for all fields L containing F;*

(c) *A is a product of separable field extensions of F.*

PROOF. (a)$\Rightarrow$(b). Let L be a field containing F. By hypothesis, there exists a field L' containing F such that $L' \otimes A$ is diagonalizable. Let L'' be a field containing (copies of) both L and L' (e.g., take L'' to be a quotient of $L \otimes L'$ by a maximal ideal). Then $L'' \otimes A = L'' \otimes_{L'} L' \otimes A$ is diagonalizable, and the map $L \otimes A \to L'' \otimes A$ defined by the inclusion $L \to L''$ is injective, and so $L \otimes A$ is reduced.

---

[1] This is Bourbaki's definition

(b)$\Rightarrow$(c). In particular, $A = A \otimes F$ is reduced, and so it is a finite product of fields (see the above discussion). Suppose that one of the factor fields F' of A is not separable over F. Then F has characteristic $p \neq 0$ and there exists an element u of F' whose minimal polynomial is of the form $g(X^p)$ with $g \in F[X]$ (see 3.6 *et seq.*). Let L be a field containing F. Then

$$L \otimes F[u] \simeq L \otimes (F[X]/(g(X^p))) \simeq L[X]/(g(X^p)).$$

If L is chosen so that the coefficients of $g(X)$ become pth powers in it, then $g(X^p)$ is a pth power in $L[X]$ (see the proof of 2.24), and so $L \otimes F[u]$ is not reduced. But $L \otimes F[u] \subset L \otimes A$, and so this contradicts the hypothesis.

(c)$\Rightarrow$(a). We may suppose that A itself is a separable field extension of F. From the primitive element theorem (5.1), we know that $A = F[u]$ for some u. Because $F[u]$ is separable over F, the minimal polynomial $f(X)$ of u is separable, which means that, in any splitting field L for f,

$$f(X) = \prod (X - u_i), \quad u_i \neq u_j \text{ for } i \neq j.$$

Now

$$L \otimes A \simeq L \otimes F[X]/(f) \simeq L[X]/(f),$$

and, according to the Chinese remainder theorem (8.1),

$$L[X]/(f) \simeq \prod_i L[X]/(X - u_i) \simeq L \times \cdots \times L. \qquad \square$$

COROLLARY 8.7 *An F-algebra A is étale if and only if $F^{\text{sep}} \otimes A$ is diagonalizable.*

PROOF. The proof that (c) implies (a) in (8.6) shows that $L \otimes A$ is diagonalizable if certain separable polynomials split in L. By definition, all separable polynomials split in F^{sep}. $\qquad \square$

COROLLARY 8.8 *Let $f \in F[X]$. Then $A = F[X]/(f)$ is an étale F-algebra if and only if f is separable.*

PROOF. Let $f = \prod_i f_i^{m_i}$ with the f_i irreducible and distinct. According to the Chinese remainder theorem (CA 2.13)

$$A \simeq \prod_i F[X]/(f_i^{m_i}).$$

The F-algebra $F[X]/(f_i^{m_i})$ is a field if and only if $m_i = 1$, in which case it is a separable extension of F if and only if f_i is separable. This completes the proof. $\qquad \square$

Not all étale F-algebras are of the form $F[X]/(f)$; for example, the F-algebra $F[X]/(f) \times F[X]/(f)$ is not.

PROPOSITION 8.9 *Finite products, tensor products, and quotients of diagonalizable (resp. étale) F-algebras are diagonalizable (resp. étale).*

PROOF. This is obvious for diagonalizable algebras, and it follows for étale algebras. $\square$

COROLLARY 8.10 *The composite of any finite set of étale F-subalgebras of an F-algebra is étale.*

PROOF. Let A be an F-algebra, and, for $i = 1,\ldots,n$, let A_i be an étale subalgebra of A. The composite $A_1 \cdots A_n$ of the A_i (i.e., the smallest F-subalgebra containing the A_i) is the image of the map

$$a_1 \otimes \cdots \otimes a_n \mapsto a_1 \cdots a_n \colon A_1 \otimes \cdots \otimes A_n \to A,$$

which is a quotient of $A_1 \otimes \cdots \otimes A_n$. $\square$

PROPOSITION 8.11 *If A is an étale F-algebra, then $F' \otimes A$ is an étale F'-algebra for any extension F' of F.*

PROOF. Let L be an extension of F such that $L \otimes A \approx L^m$, and let L' be a field containing (copies of) both L and F'. Then

$$L' \otimes_{F'} (F' \otimes A) \simeq L' \otimes A \simeq L' \otimes_L (L \otimes A) \approx L' \otimes_L L^m \simeq (L')^m. \quad \square$$

REMARK 8.12 Let A be an étale algebra over F, and write A as a product of fields, $A = \prod_i A_i$. A generator α for A as an F-algebra is a tuple (α_i) with each α_i a generator for A_i as an F-algebra. Because each A_i is separable over F, such an α exists (primitive element theorem 5.1). Choose an α, and let $f = \prod_i f_i$ be the product of the minimal polynomials of the α_i. Then f is a monic polynomial whose irreducible factors are separable.

Conversely, let f be a monic polynomial whose irreducible factors $(f_i)_i$ are separable. Then $A \overset{\text{def}}{=} \prod_i F[X]/(f_i)$ is an étale algebra over F with a canonical generator.

In this way, we get a one-to-one correspondence between the set of isomorphism classes of pairs (A, α) consisting of an étale F-algebra and a generator and the set of monic polynomials whose irreducible factors are separable.

8.13 In preparation for the next section, we review a little linear algebra. Let Ω be a Galois extension of F (possibly infinite) with Galois group G. Let V be a vector space over F, and let $V_\Omega = \Omega \otimes_F V$. Then G acts on V_Ω through its action on Ω, and the map

$$v \mapsto 1 \otimes v \colon V \to (V_\Omega)^G \stackrel{\text{def}}{=} \{v \in V_\Omega \mid \sigma v = v \text{ for all } \sigma \in G\}$$

is an isomorphism. To see this, choose an F-basis $\{e_1, \ldots, e_n\}$ for V. Then $\{e_1, \ldots, e_n\}$ is also an Ω-basis for V_Ω, and

$$\sigma(a_1 e_1 + \cdots + a_n e_n) = (\sigma a_1)e_1 + \cdots + (\sigma a_n)e_n, \quad a_i \in \Omega.$$

Therefore $a_1 e_1 + \cdots + a_n e_n$ is fixed by all $\sigma \in G$ if and only if $a_1, \ldots, a_n \in F$.

Similarly, if W is a second vector space over F, then G acts on the vector space $\mathrm{Hom}_{\Omega\text{-linear}}(V_\Omega, W_\Omega)$ by $\sigma \alpha = \sigma \circ \alpha \circ \sigma^{-1}$, and

$$\mathrm{Hom}_{F\text{-linear}}(V, W) \simeq \mathrm{Hom}_{\Omega\text{-linear}}(V_\Omega, W_\Omega)^G.$$

Again, this can be proved by choosing bases.

Classification of étale algebras over a field

We fix a separable closure Ω of F, and let $G = \mathrm{Gal}(\Omega/F)$. Recall (Chapter 7) that for every subfield E of Ω finite and Galois over F, the homomorphism

$$\sigma \mapsto \sigma|E \colon G \to \mathrm{Gal}(E/F)$$

is surjective, and its kernel is an open normal subgroup of G. Every open normal subgroup of G is of this form, and $G = \varprojlim \mathrm{Gal}(E/F)$.

By a G-**set** we mean a set S equipped with an action of G such that the map

$$G \times S \to S$$

is continuous with respect to the Krull topology on G and the discrete topology on S. This is equivalent to saying that the stabilizer of every point of S is an *open* subgroup of G. When S is finite, it is equivalent to saying that the action factors through $G \to \mathrm{Gal}(E/F)$ for some subfield E of Ω finite and Galois over F.

THE FUNCTOR $\mathcal{F}$

For an étale F-algebra A, let $\mathcal{F}(A)$ denote the set of F-algebra homomorphisms $f\colon A \to \Omega$. We let G act on $\mathcal{F}(A)$ through its action on Ω,

$$(\sigma f)(a) = \sigma(f(a)), \quad \sigma \in G,\, f \in \mathcal{F}(A),\, a \in A,$$

For some finite Galois extension E of F in Ω, the images of all homomorphism $A \to \Omega$ are contained in E,[2] and so the action of G on $\mathcal{F}(A)$ factors through $\mathrm{Gal}(E/F)$. Therefore $\mathcal{F}(A)$ is a G-set.

8.14 Let $A = F[X]/(f)$ where f is a separable polynomial in $F[X]$, and let $F[X]/(f) = F[x]$. For every homomorphism $\varphi\colon A \to \Omega$ of F-algebras, $\varphi(x)$ is a root of $f(X)$ in Ω, and the map $\varphi \mapsto \varphi(x)$ defines a one-to-one correspondence

$$\mathcal{F}(A) \xleftrightarrow{\;1:1\;} \{\text{roots of } f(X) \text{ in } \Omega\}$$

commuting with the actions of G. This is obvious from 2.1.

8.15 Let $A = A_1 \times \cdots \times A_n$ with each A_i an étale F-algebra. Because Ω is an integral domain, every homomorphism $f\colon A \to \Omega$ is zero on all but one A_i, and so, to give a homomorphism $A \to \Omega$ amounts to giving a homomorphism $A_i \to \Omega$ for some i. In other words,

$$\mathcal{F}(\textstyle\prod_i A_i) \simeq \bigsqcup_i \mathcal{F}(A_i) \quad \text{(disjoint sum)}.$$

In particular, for an étale F-algebra $A = \prod_i F_i$, F_i a field,

$$\mathcal{F}(A) \simeq \bigsqcup_i \mathrm{Hom}_F(F_i, \Omega).$$

From Proposition 2.12, we deduce that $\mathcal{F}(A)$ is finite of order $[A:F]$.

Thus, $\mathcal{F}$ is a functor from étale F-algebras to finite G-sets.

THE FUNCTOR $\mathcal{A}$

For a G-set S, we let G act on the F-algebra $\mathrm{Hom}(S, \Omega)$ of maps $S \to \Omega$ through its actions on S and Ω,

$$(\sigma f)(s) = \sigma(f(\sigma^{-1} s)), \quad \sigma \in G,\, f \in \mathrm{Hom}(S, \Omega),\, s \in S.$$

We define $\mathcal{A}(S)$ to be the set of elements of $\mathrm{Hom}(S, \Omega)$ fixed by G. Thus $\mathcal{A}(S)$ is the F-subalgebra of $\mathrm{Hom}(S, \Omega)$ consisting of the maps $f\colon S \to \Omega$ such that $f(\sigma s) = \sigma f(s)$ for all $\sigma \in G,\, s \in S$.

---

[2]Write $A = F_1 \times \cdots \times F_n$ with each F_i a field; embed each F_i in Ω, take its Galois closure, and then take the composite of the fields obtained.

8.16 Suppose that G acts transitively on S. Choose an $s \in S$, and let $H \subset G$ be its stabilizer. Then H is an open subgroup of G, and so $E = \Omega^H$ is a finite extension of F (7.13). An element f of $\mathcal{A}(S)$ is determined by its value on s, which can be any element of Ω fixed by H. It follows that the map

$$f \mapsto f(s) \colon \mathcal{A}(S) \to E$$

is an isomorphism of F-algebras.

Every element of S is of the form σs with $\sigma \in G$, and $\sigma s = \sigma' s$ if and only if $\sigma H = \sigma' H$. Similarly, every element of $\mathcal{F}(E)$ is of the form $\sigma | E$ with $\sigma \in G$, and $\sigma | E = \sigma | E'$ if and only if $\sigma H = \sigma' H$. It follows that the map

$$\sigma s \mapsto \sigma | E \colon S \to \mathcal{F}(E)$$

is an isomorphism of G-sets.

Let E be a finite separable extension E of F. Let $S = \mathcal{F}(E)$ and choose an $s \in S$, i.e., an embedding $s \colon E \hookrightarrow \Omega$. The above calculation shows that $\mathcal{A}(S) = sE$. In particular, s defines an isomorphism $E \to \mathcal{A}(\mathcal{F}(E))$.

PROPOSITION 8.17 *Let S be a finite G-set, and let $S = S_1 \sqcup \ldots \sqcup S_n$ be the decomposition of S into its G-orbits. For each i, choose an $s_i \in S_i$, and let F_i be the subfield of Ω fixed by the stabilizer of s_i.*

(a) *Each F_i is a finite separable extension of F.*

(b) *The map*

$$f \mapsto (f(s_1), \ldots, f(s_n)) \colon \mathcal{A}(S) \to F_1 \times \cdots \times F_n$$

is an isomorphism of F-algebras.

(c) *The map sending $\sigma s_i \in S_i \subset S$ to $\sigma | F_i \in \mathcal{F}(F_i) \subset \mathcal{F}(F_1 \times \cdots \times F_n)$ is an isomorphism of G-sets*

$$S \to \mathcal{F}(F_1 \times \cdots \times F_n).$$

PROOF. This follows directly from the special case considered in 8.16. $\square$

PROPOSITION 8.18 *For every finite G-set S, the F-algebra $\mathcal{A}(S)$ is étale with degree equal to $|S|$. Moreover, every étale F-algebra A is of the form $\mathcal{A}(S)$ for some G-set S. More precisely,*

$$A \simeq \mathcal{A}(\mathcal{F}(A)).$$

PROOF. The first statement follows from (8.6) and (8.17). We prove the third statement. There is a canonical isomorphism of Ω-algebras

$$a \otimes c \mapsto (\sigma a \cdot c)_{\sigma \in \mathcal{F}(A)} \colon \Omega \otimes A \to \prod_{\sigma \in \mathcal{F}(A)} \Omega.$$

When we let G act on $\Omega \otimes A$ through Ω, and pass to the fixed elements, we obtain an isomorphism

$$A \overset{8.13}{=} (\Omega \otimes A)^G \simeq \mathcal{A}(\mathcal{F}(A)).$$

This implies the second statement of the proposition (which can also be deduced from 8.17). $\qquad\square$

PROPOSITION 8.19 *Let S be a finite G-set. An element $s \in S$ defines a homomorphism of F-algebras $f \mapsto f(s) \colon \mathcal{A}(S) \to \Omega$, and every homomorphism of F-algebras $\mathcal{A}(S) \to \Omega$ is of this form for a unique s. Thus $S \simeq \mathcal{F}(\mathcal{A}(S))$.*

PROOF. We leave this as an exercise. $\qquad\square$

PROPOSITION 8.20 *For all étale F-algebras A and B, the map*

$$\mathrm{Hom}_{F\text{-algebras}}(A, B) \to \mathrm{Hom}_{G\text{-sets}}(\mathcal{F}(B), \mathcal{F}(A))$$

defined by $\mathcal{F}$ is bijective.

PROOF. Let A and B be étale F-algebras. Under the isomorphism

$$\mathrm{Hom}_{F\text{-linear}}(A, B) \overset{8.13}{\simeq} \mathrm{Hom}_{\Omega\text{-linear}}(A_\Omega, B_\Omega)^G,$$

F-algebra homomorphisms correspond to Ω-algebra homomorphisms, and so

$$\mathrm{Hom}_{F\text{-algebra}}(A, B) \simeq \mathrm{Hom}_{\Omega\text{-algebra}}(A_\Omega, B_\Omega)^G.$$

From Corollary 8.7, we know that A_Ω (resp. B_Ω) is a product of copies of Ω indexed by the elements of $\mathcal{F}(A)$ (resp. $\mathcal{F}(B)$). Let t be a map of sets $\mathcal{F}(B) \to \mathcal{F}(A)$. Then

$$(a_i)_{i \in \mathcal{F}(A)} \mapsto (b_j)_{j \in \mathcal{F}(B)}, \quad b_j = a_{t(j)},$$

is a homomorphism of Ω-algebras $A_\Omega \to B_\Omega$, and every homomorphism of Ω-algebras $A_\Omega \to B_\Omega$ is of this form for a unique t. Thus

$$\mathrm{Hom}_{\Omega\text{-algebra}}(A_\Omega, B_\Omega) \simeq \mathrm{Hom}_{\mathrm{Sets}}(\mathcal{F}(B), \mathcal{F}(A)).$$

This isomorphism is compatible with the actions of G, and so

$$\mathrm{Hom}_{\Omega\text{-algebra}}(A_\Omega, B_\Omega)^G \simeq \mathrm{Hom}_{\mathrm{Sets}}(\mathcal{F}(B), \mathcal{F}(A))^G.$$

In other words,

$$\mathrm{Hom}_{F\text{-algebra}}(A, B) \simeq \mathrm{Hom}_{G\text{-sets}}(\mathcal{F}(B), \mathcal{F}(A)). \qquad \square$$

THEOREM 8.21 *The functor $A \rightsquigarrow \mathcal{F}(A)$ is a contravariant equivalence from the category of étale F-algebras to the category of finite G-sets with quasi-inverse $\mathcal{A}$.*

PROOF. This summarizes the results in the last three propositions. $\qquad \square$

It is possible to prove Theorem 8.21 directly, without using Galois theory, and then deduce Galois theory from it.

GENERALIZATION OF THEOREM 8.21

Let Ω be a Galois extension of F (finite or infinite), and let $G = \mathrm{Gal}(\Omega/F)$. An étale F-algebra A is **split** by Ω if $\Omega \otimes A$ is isomorphic to a product of copies of Ω. For such an F-algebra, we let $\mathcal{F}(A) = \mathrm{Hom}_{k\text{-algebra}}(A, \Omega)$.

THEOREM 8.22 *The functor $A \rightsquigarrow \mathcal{F}(A)$ is a contravariant equivalence from the category of étale F-algebras split by Ω to the category of finite G-sets.*

When Ω is a finite extension of F, the continuity condition for G-sets can be omitted.

The proof of Theorem 8.22 is the same as that of Theorem 8.21. Alternatively, deduce it from 8.21 by noting that the categories in question are the full subcategories of the categories in 8.21 whose objects are those on which $\mathrm{Gal}(F^{\mathrm{sep}}/\Omega$ acts trivially.

GEOMETRIC RE-STATEMENT OF THEOREM 8.21

In this subsection, we assume that the reader is familiar with the notion of an algebraic variety over a field F (geometrically reduced separated scheme of finite type over F). The functor $A \rightsquigarrow \mathrm{Spec}(A)$ is a contravariant equivalence from the category of étale algebras over F to the category of zero-dimensional algebraic varieties over F. In particular, all zero-dimensional algebraic varieties are affine. If $V = \mathrm{Spec}(A)$, then

$$\mathrm{Hom}_{F\text{-algebra}}(A, \Omega) \simeq \mathrm{Hom}_{\mathrm{Spec}(F)}(\mathrm{Spec}(\Omega), V) \stackrel{\mathrm{def}}{=} V(\Omega)$$

(set of points of V with coordinates in Ω).

THEOREM 8.23 *The functor $V \rightsquigarrow V(\Omega)$ is an equivalence from the category of zero-dimensional algebraic varieties over F to the category of finite continuous G-sets. Under this equivalence, connected varieties correspond to sets with a transitive action.*

PROOF. Combine Theorem 8.21 with the equivalence $A \rightsquigarrow \operatorname{Spec}(A)$. $\square$

Comparison with the theory of covering spaces.

The reader should compare (8.21) and (8.23) with the following statement:

> Let X be a connected and locally simply connected topological space. Let $x \in X$, and let $\pi_1(X, x)$ be the fundament group (homotopy classes of loops based at x). Let $Y \to X$ be a covering space, and let $\mathcal{F}(Y)$ denote the preimage of x in Y. There is a natural action of $\pi_1(X, x)$ on $\mathcal{F}(Y)$: let γ be a (small) loop based at x regarded as a function $\gamma \colon [0, 1] \to X$, and let $y \in \mathcal{F}(Y)$; then γ it lifts to a function $\gamma_y \colon [0, 1] \to Y$ such that $\gamma_y(0) = y$, and we define $\gamma \cdot y = \gamma_y(1)$. The functor $E \rightsquigarrow \mathcal{F}(E)$ is an equivalence from the category of covering spaces of F to the category of finite $\pi_1(X, x)$-sets.

For more on this, see the section on the étale fundamental group in my notes *Lectures on Étale Cohomology* or Szamuely, *Galois groups and fundamental groups*, CUP, 2009.

ASIDE 8.24 (FOR THE EXPERTS) It is possible to define the "absolute Galois group" of a field F canonically and without assuming the axiom of choice. Let $\mathcal{S}$ denote the category of sheaves of $\mathbb{Q}$-vector spaces on $\operatorname{Spec}(F)_{\mathrm{et}}$ having the property that $S(A)$ is a finite-dimensional vector space for all A and the dimension of $S(K)$, K a field, is bounded. This is a tannakian category, and we define the absolute Galois group π of F to be the fundamental group of this category. This is an affine group scheme in the category $\mathcal{S}$. For any choice of a separable closure F^{sep} of F, we get a fibre functor ω on $\mathcal{S}$, and $\omega(\pi) = \operatorname{Gal}(F^{\mathrm{sep}}/F)$. See Julian Rosen, *A choice-free absolute Galois group and Artin motives*, arXiv:1706.06573.

CHAPTER 9

Transcendental Extensions

In this chapter we consider fields $\Omega \supset F$ with Ω much bigger than F. For example, we could have $\mathbb{C} \supset \mathbb{Q}$.

Algebraic independence

Elements $\alpha_1, \ldots, \alpha_n$ of Ω give rise to an F-homomorphism

$$f \mapsto f(\alpha_1, \ldots, \alpha_n) \colon F[X_1, \ldots, X_n] \to \Omega.$$

If the kernel of this homomorphism is zero, then the α_i are said to be **algebraically independent** over F, and otherwise, they are **algebraically dependent** over F. Thus, the α_i are algebraically dependent over F if there exists a nonzero polynomial $f(X_1, \ldots, X_n) \in F[X_1, \ldots, X_n]$ such that $f(\alpha_1, \ldots, \alpha_n) = 0$, and they are algebraically independent if

$$a_{i_1, \ldots, i_n} \in F, \quad \sum a_{i_1, \ldots, i_n} \alpha_1^{i_1} \ldots \alpha_n^{i_n} = 0 \implies a_{i_1, \ldots, i_n} = 0 \text{ all } i_1, \ldots, i_n.$$

Note the similarity with linear independence. In fact, if f is required to be homogeneous of degree 1, then the definition becomes that of linear independence.

EXAMPLE 9.1 (a) A single element α is algebraically independent over F if and only if it is transcendental over F.

(b) The complex numbers π and e are certainly expected to be algebraically independent over $\mathbb{Q}$, but this has not been proved.

An infinite set A is **algebraically independent** over F if every finite subset of A is algebraically independent; otherwise, it is **algebraically dependent** over F.

REMARK 9.2 If $\alpha_1, \dots, \alpha_n$ are algebraically independent over F, then the map

$$f(X_1, \dots, X_n) \mapsto f(\alpha_1, \dots, \alpha_n) \colon F[X_1, \dots, X_n] \to F[\alpha_1, \dots, \alpha_n]$$

is injective, and hence an isomorphism. This isomorphism then extends to the fields of fractions,

$$X_i \mapsto \alpha_i \colon F(X_1, \dots, X_n) \to F(\alpha_1, \dots, \alpha_n)$$

In this case, $F(\alpha_1, \dots, \alpha_n)$ is called a **pure transcendental extension** of F. The polynomial

$$f(X) = X^n - \alpha_1 X^{n-1} + \cdots + (-1)^n \alpha_n$$

has Galois group S_n over $F(\alpha_1, \dots, \alpha_n)$ (see 5.40).

LEMMA 9.3 *Let $\gamma \in \Omega$ and let $A \subset \Omega$. The following conditions are equivalent:*

(a) *γ is algebraic over $F(A)$;*

(b) *there exist $\beta_1, \dots, \beta_n \in F(A)$ such that $\gamma^n + \beta_1 \gamma^{n-1} + \cdots + \beta_n = 0$;*

(c) *there exist $\beta_0, \beta_1, \dots, \beta_n \in F[A]$, not all 0, such that $\beta_0 \gamma^n + \beta_1 \gamma^{n-1} + \cdots + \beta_n = 0$;*

(d) *there exists an $f(X_1, \dots, X_m, Y) \in F[X_1 \dots, X_m, Y]$ and $\alpha_1, \dots, \alpha_m \in A$ such that $f(\alpha_1, \dots, \alpha_m, Y) \neq 0$ but $f(\alpha_1, \dots, \alpha_m, \gamma) = 0$.*

PROOF. (a) $\Longrightarrow$ (b) $\Longrightarrow$ (c) $\Longrightarrow$ (a) are obvious.

(d) $\Longrightarrow$ (c). Write $f(X_1, \dots, X_m, Y)$ as a polynomial in Y with coefficients in the ring $F[X_1, \dots, X_m]$,

$$f(X_1, \dots, X_m, Y) = \sum f_i(X_1, \dots, X_m) Y^{n-i}.$$

Then (c) holds with $\beta_i = f_i(\alpha_1, \dots, \alpha_m)$.

(c) $\Longrightarrow$ (d). The β_i in (c) can be expressed as polynomials in a finite number of elements $\alpha_1, \dots, \alpha_m$ of A, say, $\beta_i = f_i(\alpha_1, \dots, \alpha_m)$ with $f_i \in F[X_1, \dots, X_m]$. Then (d) holds with $f = \sum f_i(X_1, \dots, X_m) Y^{n-i}$. $\qquad \square$

DEFINITION 9.4 When γ satisfies the equivalent conditions of Lemma 9.3, it is said to be **algebraically dependent** on A (over F). A set B is **algebraically dependent** on A if every element of B is algebraically dependent on A.

The theory in the remainder of this chapter is logically very similar to a part of linear algebra. It is useful to keep the following correspondences in mind:

Linear algebra	Transcendence
linearly independent	algebraically independent
$A \subset \mathrm{span}(B)$	A algebraically dependent on B
basis	transcendence basis
dimension	transcendence degree

Transcendence bases

THEOREM 9.5 (FUNDAMENTAL RESULT) *Let* $A = \{\alpha_1, ..., \alpha_m\}$ *and* $B = \{\beta_1, ..., \beta_n\}$ *be two subsets of* Ω. *Assume*

   (a) *A is algebraically independent (over F);*

   (b) *A is algebraically dependent on B (over F).*

Then $m \le n$.

We first prove two lemmas.

LEMMA 9.6 (THE EXCHANGE PROPERTY) *Let $\{\alpha_1, ..., \alpha_m\}$ be a subset of Ω; if β is algebraically dependent on $\{\alpha_1, ..., \alpha_m\}$ but not on $\{\alpha_1, ..., \alpha_{m-1}\}$, then α_m is algebraically dependent on $\{\alpha_1, ..., \alpha_{m-1}, \beta\}$.*

PROOF. Because β is algebraically dependent on $\{\alpha_1, ..., \alpha_m\}$, there exists a polynomial $f(X_1, ..., X_m, Y)$ with coefficients in F such that

$$f(\alpha_1, ..., \alpha_m, Y) \ne 0, \quad f(\alpha_1, ..., \alpha_m, \beta) = 0.$$

Write f as a polynomial in X_m,

$$f(X_1, ..., X_m, Y) = \sum_i a_i(X_1, ..., X_{m-1}, Y) X_m^{n-i},$$

and observe that, because $f(\alpha_1, ..., \alpha_m, Y) \ne 0$, at least one of the polynomials

$$a_i(\alpha_1, ..., \alpha_{m-1}, Y),$$

say a_{i_0}, is not the zero polynomial. Because β is not algebraically dependent on

$$\{\alpha_1, ..., \alpha_{m-1}\},$$

$a_{i_0}(\alpha_1, ..., \alpha_{m-1}, \beta) \neq 0$. Therefore, $f(\alpha_1, ..., \alpha_{m-1}, X_m, \beta) \neq 0$. Because $f(\alpha_1, ..., \alpha_m, \beta) = 0$, this shows that α_m is algebraically dependent on the set $\{\alpha_1, ..., \alpha_{m-1}, \beta\}$. $\qquad\square$

LEMMA 9.7 (TRANSITIVITY OF ALGEBRAIC DEPENDENCE) *If C is algebraically dependent on B, and B is algebraically dependent on A, then C is algebraically dependent on A.*

PROOF. The argument in the proof of Proposition 1.45 shows that if γ is algebraic over a field E which is algebraic over a field F, then γ is algebraic over F (if $a_1, \ldots, a_n$ are the coefficients of the minimal polynomial of γ over E, then the field $F[a_1, \ldots, a_n, \gamma]$ has finite degree over F). Apply this with $E = F(A \cup B)$ and $F = F(A)$. $\qquad\square$

PROOF (OF THEOREM 9.5) Let k be the number of elements that A and B have in common. If $k = m$, then $A \subset B$, and certainly $m \leq n$. Suppose that $k < m$, and write $B = \{\alpha_1, ..., \alpha_k, \beta_{k+1}, ..., \beta_n\}$. Since α_{k+1} is algebraically dependent on $\{\alpha_1, ..., \alpha_k, \beta_{k+1}, ..., \beta_n\}$ but not on $\{\alpha_1, ..., \alpha_k\}$, there will be a β_j, $k + 1 \leq j \leq n$, such that α_{k+1} is algebraically dependent on $\{\alpha_1, ..., \alpha_k, \beta_{k+1}, ..., \beta_j\}$ but not

$$\{\alpha_1, ..., \alpha_k, \beta_{k+1}, ..., \beta_{j-1}\}.$$

The exchange lemma then shows that β_j is algebraically dependent on

$$B_1 \overset{\text{def}}{=} B \cup \{\alpha_{k+1}\} \smallsetminus \{\beta_j\}.$$

Therefore B is algebraically dependent on B_1, and so A is algebraically dependent on B_1 (by 9.7). If $k + 1 < m$, repeat the argument with A and B_1. Eventually we'll achieve $k = m$, and $m \leq n$. $\qquad\square$

DEFINITION 9.8 A ***transcendence basis*** for Ω over F is an algebraically independent set A such that Ω is algebraic over $F(A)$.

LEMMA 9.9 *If Ω is algebraic over $F(A)$, and A is minimal among subsets of Ω with this property, then it is a transcendence basis for Ω over F.*

PROOF. If A is not algebraically independent, then there is an $\alpha \in A$ that is algebraically dependent on $A \smallsetminus \{\alpha\}$. It follows from Lemma 9.7 that Ω is algebraic over $F(A \smallsetminus \{\alpha\})$. $\qquad\square$

THEOREM 9.10 *If there is a finite subset $A \subset \Omega$ such that Ω is algebraic over $F(A)$, then Ω has a finite transcendence basis over F. Moreover, every transcendence basis is finite, and they all have the same number of elements.*

PROOF. In fact, every minimal subset A' of A such that Ω is algebraic over $F(A')$ will be a transcendence basis. The second statement follows from Theorem 9.5. $\square$

LEMMA 9.11 *Suppose that A is algebraically independent, but that $A \cup \{\beta\}$ is algebraically dependent. Then β is algebraic over $F(A)$.*

PROOF. The hypothesis is that there exists a nonzero polynomial

$$f(X_1, ..., X_n, Y) \in F[X_1, ..., X_n, Y]$$

such that $f(\alpha_1, ..., \alpha_n, \beta) = 0$, some distinct $\alpha_1, ..., \alpha_n \in A$. Because A is algebraically independent, Y does occur in f. Therefore

$$f = g_0 Y^m + g_1 Y^{m-1} + \cdots + g_m, \quad g_i \in F[X_1, ..., X_n], \quad g_0 \neq 0, \quad m \geq 1.$$

As $g_0 \neq 0$ and the α_i are algebraically independent, $g_0(\alpha_1, ..., \alpha_n) \neq 0$. Because β is a root of

$$f = g_0(\alpha_1, ..., \alpha_n)X^m + g_1(\alpha_1, ..., \alpha_n)X^{m-1} + \cdots + g_m(\alpha_1, ..., \alpha_n),$$

it is algebraic over $F(\alpha_1, ..., \alpha_n) \subset F(A)$. $\square$

PROPOSITION 9.12 *Every maximal algebraically independent subset of Ω is a transcendence basis for Ω over F.*

PROOF. We have to prove that Ω is algebraic over $F(A)$ if A is maximal among algebraically independent subsets. But the maximality implies that, for every $\beta \in \Omega \smallsetminus A$, $A \cup \{\beta\}$ is algebraically dependent, and so the lemma shows that β is algebraic over $F(A)$. $\square$

We now need to assume Zorn's lemma.

THEOREM 9.13 *Every algebraically independent subset S of Ω is contained in a transcendence basis for Ω over F; in particular, transcendence bases exist.*

PROOF. Let $\mathcal{S}$ be the set of algebraically independent subsets of Ω containing S, partially ordered by inclusion. Let T be a totally ordered subset of $\mathcal{S}$, and let $B = \bigcup\{A \mid A \in T\}$. I claim that $B \in \mathcal{S}$, i.e., that B is algebraically independent. If not, there exists a finite subset B' of B that is not algebraically independent. But such a subset will be contained in one of the sets in T, which is a contradiction. Now Zorn's lemma shows that there exists a maximal algebraically independent set containing S, which Proposition 9.12 shows to be a transcendence basis for Ω over F. $\qquad\square$

It is possible to show that any two (possibly infinite) transcendence bases for Ω over F have the same cardinality. The cardinality of a transcendence basis for Ω over F is called the **transcendence degree** of Ω over F. For example, the pure transcendental extension $F(X_1,\ldots,X_n)$ has transcendence degree n over F.

EXAMPLE 9.14 Let $p_1,\ldots,p_n$ be the elementary symmetric polynomials in $X_1,\ldots,X_n$. The field $F(X_1,\ldots,X_n)$ is algebraic over $F(p_1,\ldots,p_n)$, and so $\{p_1,p_2,\ldots,p_n\}$ contains a transcendence basis for $F(X_1,\ldots,X_n)$. Because $F(X_1,\ldots,X_n)$ has transcendence degree n, the p_i's must themselves be a transcendence basis.

EXAMPLE 9.15 Let Ω be the field of meromorphic functions on a compact complex manifold M.

(a) The only meromorphic functions on the Riemann sphere are the rational functions in z. Hence, in this case, Ω is a pure transcendental extension of $\mathbb{C}$ of transcendence degree 1.

(b) If M is a Riemann surface, then the transcendence degree of Ω over $\mathbb{C}$ is 1, and Ω is a pure transcendental extension of $\mathbb{C} \iff M$ is isomorphic to the Riemann sphere

(c) If M has complex dimension n, then the transcendence degree is $\leq n$, with equality holding if M is embeddable in some projective space.

PROPOSITION 9.16 *Any two algebraically closed fields with the same transcendence degree over F are F-isomorphic.*

PROOF. Choose transcendence bases A and A' for the two fields. By assumption, there exists a bijection $A \to A'$, which defines an F-isomorphism $F[A] \to F[A']$, and hence an F-isomorphism of the fields of fractions $F(A) \to F(A')$. Use this isomorphism to identify $F(A)$ with $F(A')$. Then the two fields in question are algebraic closures of the same field, and hence are isomorphic (Theorem 6.6). $\qquad\square$

REMARK 9.17 Any two algebraically closed fields with the same uncountable cardinality and the same characteristic are isomorphic. The idea of the proof is as follows. Let F and F' be the prime subfields of Ω and Ω'; we can identify F with F'. Then show that when Ω is uncountable, the cardinality of Ω is the same as the cardinality of a transcendence basis over F. Finally, apply the proposition.

REMARK 9.18 What are the automorphisms of $\mathbb{C}$? There are only two continuous automorphisms (cf. Exercise A-8 and solution). When we assume Zorn's lemma, it is easy to construct many: choose a transcendence basis A for $\mathbb{C}$ over $\mathbb{Q}$, and choose a permutation α of A; then α defines an isomorphism $\mathbb{Q}(A) \to \mathbb{Q}(A)$, which can be extended to an automorphism of $\mathbb{C}$. Without Zorn's lemma, there are only two, because the noncontinuous automorphisms are nonmeasurable,[1] and it is known that the Zorn's lemma is required to construct nonmeasurable functions.[2]

Lüroth's theorem

THEOREM 9.19 (LÜROTH) *Let $L = F(X)$ with X transcendental over F. Every subfield E of L properly containing F is of the form $E = F(u)$ for some $u \in L$ transcendental over F.*

We first sketch a geometric proof of Lüroth's theorem. The inclusion of E into L corresponds to a map from the projective line $\mathbb{P}^1$ onto a complete regular curve C. The Riemann-Hurwitz formula shows that C has genus 0. Since it has an F-rational point (the image of any F-rational point of $\mathbb{P}^1$), it is isomorphic to $\mathbb{P}^1$. Therefore $E = F(u)$ for some $u \in L$ transcendental over F.

Before giving the elementary proof, we review Gauss's lemma and its consequences.

GAUSS'S LEMMA

Let R be a unique factorization domain, and let Q be its field of fractions, for example, $R = F[X]$ and $Q = F(X)$. A polynomial $f(T) = \sum a_i T^i$ in

---

[1]A fairly elementary theorem of G. Mackey says that every measurable homomorphism from a locally compact group to a topological group is continuous if both groups are second countable. See Theorem B.3, p. 198 of Zimmer, *Ergodic theory and semisimple groups*, 1984.

[2]Solovay, *A model of set-theory in which every set of reals is Lebesgue measurable*. Ann. of Math. (2) 92 (1970), 1–56.

$R[T]$ is said to be **primitive** if its coefficients a_i have no common factor other than units. Every polynomial f in $Q[X]$ can be written $f = c(f) \cdot f_1$ with $c(f) \in Q$ and f_1 primitive (write $f = af/a$ with a a common denominator for the coefficients of f, and then write $f = (b/a)f_1$ with b the greatest common divisor of the coefficients of af). The element $c(f)$ is uniquely determined up to a unit, and $f \in R[X]$ if and only if $c(f) \in R$.

9.20 *If $f, g \in R[T]$ are primitive, so also is fg.*

Let $f = \sum a_i T^i$ and $g = \sum b_i T^i$, and let p be a prime element of R. Because f is primitive, there exists a coefficient a_i not divisible by p — let a_{i_1} be the first such coefficient. Similarly, let b_{i_2} be the first coefficient of g not divisible by p. Then the coefficient of $T^{i_1 + i_2}$ in fg is not divisible by p. This shows that fg is primitive.

9.21 *For any $f, g \in R[T]$, $c(fg) = c(f)c(g)$ and $(fg)_1 = f_1 g_1$.*

Let $f = c(f)f_1$ and $g = c(g)g_1$ with f_1 and g_1 primitive. Then $fg = c(f)c(g)f_1 g_1$ with $f_1 g_1$ primitive, and so $c(fg) = c(f)c(g)$ and $(fg)_1 = f_1 g_1$.

9.22 *Let f be a polynomial in $R[T]$. If f factors into the product of two nonconstant polynomials in $Q[T]$, then it factors into the product of two nonconstant polynomials in $R[T]$.*

Suppose that $f = gh$ in $Q[T]$. Then $f_1 = g_1 h_1$ in $R[T]$, and $f = c(f) \cdot f_1 = (c(f) \cdot g_1)h_1$ is a factorization of f in $R[T]$.

9.23 *Let $f, g \in R[T]$. If f divides g in $Q[T]$ and f is primitive, then it divides g in $R[T]$.*

Let $fq = g$ with $q \in Q[T]$. Then $c(q) = c(g) \in R$, and so $q \in R[T]$.

PROOF OF LÜROTH'S THEOREM

We define the degree $\deg(u)$ of an element u of $F(X)$ to be the larger of the degrees of the numerator and denominator of u when it is expressed in its simplest form.

LEMMA 9.24 *Let $u \in F(X) \smallsetminus F$. Then u is transcendental over F, X is algebraic over $F(u)$, and $[F(X) : F(u)] = \deg(u)$.*

PROOF. Let $u(X) = a(X)/b(X)$ with $a(X)$ and $b(X)$ relatively prime polynomials. Then $a(T) - b(T)u$ is a polynomial in $F(u)[T]$ having X as a root, and so X is algebraic over $F(u)$. It follows that u is transcendental over F (else X would be algebraic over F; 1.31).

The polynomial $a(T) - b(T)Z \in F[Z,T]$ is clearly irreducible. As u is transcendental over F,

$$F[Z,T] \simeq F[u,T], \quad Z \leftrightarrow u, \quad T \leftrightarrow T,$$

and so $a(T) - b(T)u$ is irreducible in $F[u,T]$, and hence also in $F(u)[T]$ by Gauss's lemma (9.22). It has X as a root, and so, up to a constant, it is the minimal polynomial of X over $F(u)$, and its degree is $\deg(u)$, which proves the lemma. $\qquad\square$

EXAMPLE 9.25 We have $F(X) = F(u)$ if and only if

$$u = \frac{aX + b}{cX + d}$$

with $ad - bc \neq 0$.

We now prove Theorem 9.19. For any $u \in E \smallsetminus F$,

$$[F(X):E] \leq [F(X):F(u)] = \deg(u),$$

and so X is algebraic over E. Let

$$f(T) = T^n + a_1 T^{n-1} + \cdots + a_n, \quad a_i \in E, \quad n = [F(X):E],$$

be its minimal polynomial. As X is transcendental over F, some $a_j \notin F$, and we'll show that $E = F(a_j)$ for such an a_j.

Let $d(X) \in F[X]$ be a polynomial of least degree such that $d(X)a_i(X) \in F[X]$ for all i, and let

$$f_1(X,T) = df(T) = dT^n + da_1 T^{n-1} + \cdots + da_n \in F[X,T].$$

Then f_1 is primitive as a polynomial in T, i.e., $\gcd(d, da_1, \ldots, da_n) = 1$ in $F[X]$. The degree m of f_1 in X is the largest degree of one of the polynomials $da_1, da_2, \ldots$, say,

$$m = \deg(da_i).$$

Write $a_i = b/c$ with b, c relatively prime polynomials in $F[X]$. Now $b(T) - c(T)a_i(X)$ is a polynomial in $E[T]$ having X as a root, and so it is divisible by f, say

$$f(T) \cdot q(T) = b(T) - c(T) \cdot a_i(X), \quad q(T) \in E[T].$$

On multiplying through by $c(X)$, we find that

$$c(X) \cdot f(T) \cdot q(T) = c(X) \cdot b(T) - c(T) \cdot b(X).$$

As f_1 differs from f by a nonzero element of $F(X)$, the equation shows that f_1 divides $c(X) \cdot b(T) - c(T) \cdot b(X)$ in $F(X)[T]$, but f_1 is primitive in $F[X][T]$, and so it divides the polynomial in $F[X][T] = F[X,T]$ (by 9.23), i.e., there exists a polynomial $h \in F[X,T]$ such that

$$f_1(X,T) \cdot h(X,T) = c(X) \cdot b(T) - c(T) \cdot b(X). \tag{14}$$

In (14), the polynomial $c(X) \cdot b(T) - c(T) \cdot b(X)$ has degree at most m in X, and m is the degree of $f_1(X,T)$ in X. Therefore, $c(X) \cdot b(T) - c(T) \cdot b(X)$ has degree exactly m in X, and $h(X,T)$ has degree 0 in X, i.e., $h \in F[T]$. It now follows from (14) that $c(X) \cdot b(T) - c(T) \cdot b(X)$ is not divisible by a nonconstant polynomial in $F[X]$.

The polynomial $c(X) \cdot b(T) - c(T) \cdot b(X)$ is symmetric in X and T, i.e., it is unchanged when they are swapped. Therefore, it has degree m in T and it is not divisible by a nonconstant polynomial in $F[T]$. It now follows from (14) that h is not divisible by a nonconstant polynomial in $F[T]$, and so it lies in $F^\times$. We conclude that $f_1(X,T)$ is a constant multiple of $c(X) \cdot b(T) - c(T) \cdot b(X)$.

On comparing degrees in T in (14), we see that $n = m$. Thus

$$[F(X):F(a_i)] \overset{9.24}{=} \deg(a_i) \le \deg(da_i) = m = n = [F(X):E] \le [F(X):F(a_i)].$$

Hence, equality holds throughout, and so $E = F[a_i]$.

Finally, if $a_j \notin F$, then

$$[F(X):E] \le [F(X):F(a_j)] \overset{9.24}{=} \deg(a_j) \le \deg(da_j) \le \deg(da_i) = [F(X):E],$$

and so $E = F(a_j)$ as claimed.

REMARK 9.26 Lüroth's theorem fails when there is more than one variable — see Zariski's example (footnote to 5.5) and Swan's example (Remark 5.41). However, the following is true: if $[F(X,Y):E] < \infty$ and F is algebraically closed of characteristic zero, then E is a pure transcendental extension of F (Theorem of Zariski, 1958).

NOTES Lüroth proved his theorem over $\mathbb{C}$ in 1876. For general fields, it was proved by Steinitz in 1910 by the above argument.

Separating transcendence bases

Let $E \supset F$ be fields with E finitely generated over F. A subset $\{x_1,\ldots,x_d\}$ of E is a **separating transcendence basis** for E/F if it is algebraically independent over F and E is a finite *separable* extension of $F(x_1,\ldots,x_d)$.

THEOREM 9.27 *If F is perfect, then every finitely generated extension E of F admits a separating transcendence basis over F.*

PROOF. If F has characteristic zero, then every transcendence basis is separating, and so the statement becomes that of Theorem 9.10. Thus, we may assume F has characteristic $p \neq 0$. Because F is perfect, every polynomial in $X_1^p,\ldots,X_n^p$ with coefficients in F is a pth power in $F[X_1,\ldots,X_n]$:

$$\sum a_{i_1\cdots i_n} X_1^{i_1 p} \ldots X_n^{i_n p} = \left(\sum a_{i_1\cdots i_n}^{\frac{1}{p}} X_1^{i_1} \ldots X_n^{i_n} \right)^p.$$

Let $E = F(x_1,\ldots,x_n)$, and assume that $n > d + 1$, where d is the transcendence degree of E over F. After renumbering, we may suppose that $x_1,\ldots,x_d$ are algebraically independent (9.9). Then $f(x_1,\ldots,x_{d+1}) = 0$ for some nonzero irreducible polynomial $f(X_1,\ldots,X_{d+1})$ with coefficients in F. Not all $\partial f/\partial X_i$ are zero, for otherwise f would be a polynomial in $X_1^p,\ldots,X_{d+1}^p$, which implies that it is a pth power. After renumbering $x_1,\ldots,x_{d+1}$, we may suppose that $\partial f/\partial X_{d+1} \neq 0$. Then x_{d+1} is separably algebraic over $F(x_1,\ldots,x_d)$ and $F(x_1,\ldots,x_{d+1},x_{d+2})$ is algebraic over $F(x_1,\ldots,x_{d+1})$, hence over $F(x_1,\ldots,x_d)$ (1.31), and so, by the primitive element theorem (5.1), there is an element y such that $F(x_1,\ldots,x_{d+2}) = F(x_1,\ldots,x_d,y)$. Thus E is generated by $n-1$ elements (as a field containing F). After repeating the process, possibly several times, we will have $E = F(z_1,\ldots,z_{d+1})$ with z_{d+1} separable over $F(z_1,\ldots,z_d)$.□

ASIDE 9.28 In fact, we showed that E admits a separating transcendence basis with $d + 1$ elements where d is the transcendence degree. This has the following geometric interpretation: every irreducible algebraic variety of dimension d over a perfect field F is birationally equivalent with a hypersurface H in $\mathbb{A}^{d+1}$ for which the projection $(a_1,\ldots,a_{d+1}) \mapsto (a_1,\ldots,a_d)$ realizes $F(H)$ as a finite separable extension of $F(\mathbb{A}^d)$ (see my notes *Algebraic Geometry*).

Transcendental Galois theory

THEOREM 9.29 *Let Ω be an algebraically closed field and let F be a perfect subfield of Ω. If $\alpha \in \Omega$ is fixed by all F-automorphisms of Ω, then $\alpha \in F$, i.e., $\Omega^{\mathrm{Aut}(\Omega/F)} = F$.*

PROOF. Let $\alpha \in \Omega \smallsetminus F$. If α is algebraic over F, then there is an F-homomorphism $F[\alpha] \to \Omega$ sending α to a conjugate of α in Ω different from α. This homomorphism extends to an isomorphism $F^{\mathrm{al}} \to F^{\mathrm{al}} \subset \Omega$, where F^{al} is the algebraic closure of F in Ω (by 6.6). Now choose a transcendence basis A for Ω over F^{al}. We can extend our isomorphism to an isomorphism $F^{\mathrm{al}}(A) \to F^{\mathrm{al}}(A) \subset \Omega$ by mapping each element of A to itself. Finally, we can extend this isomorphism to an isomorphism from the algebraic closure Ω of $F^{\mathrm{al}}(A)$ to Ω.

If α is transcendental over F, then it is part of a transcendence basis A for Ω over F (see 9.13). If $A \neq \{\alpha\}$, then there exists an automorphism σ of A such that $\sigma(\alpha) \neq \alpha$. Now σ defines an F-homomorphism $F(A) \to \Omega$, which extends to an isomorphism $\Omega \to \Omega$ as before. If $A = \{\alpha\}$, then we let $F(\alpha) \to \Omega$ be the F-homomorphism sending α to $\alpha + 1$. Again, this extends to an isomorphism $\Omega \to \Omega$. $\qquad\square$

Let $\Omega \supset F$ be fields and let $G = \mathrm{Aut}(\Omega/F)$. For any finite subset S of Ω, let
$$G(S) = \{\sigma \in G \mid \sigma s = s \text{ for all } s \in S\}.$$

Then, as in §7, the subgroups $G(S)$ of G form a neighbourhood base at the identity for a topology on the group G, which we again call the **Krull topology**. The same argument as in §7 shows that this topology is Hausdorff (but it is not necessarily compact).

THEOREM 9.31 *Let $\Omega \supset F$ be fields such that $\Omega^{\mathrm{Aut}(\Omega/F)} = F$, and let $G = \mathrm{Aut}(\Omega/F)$.*

(a) For every finite extension E of F in Ω, $\Omega^{\mathrm{Aut}(\Omega/E)} = E$.

(b) The maps
$$H \mapsto \Omega^H, \quad M \mapsto \mathrm{Aut}(\Omega/M) \tag{15}$$

are inverse bijections between the set of compact subgroups of G and the set of intermediate fields over which Ω is Galois (possibly infinite):

$$\{\text{compact subgroups of } G\} \leftrightarrow \{\text{fields } M \text{ such that } F \subset M \stackrel{\text{Galois}}{\subset} \Omega\}.$$

(c) If there exists an intermediate field M finitely generated over F such that Ω is Galois over M, then G is locally compact, and under (15), the open compact subgroups of G correspond to such M.

(d) Let H be a subgroup of G, and let $M = \Omega^H$. Then the algebraic closure M_1 of M is Galois over M. If moreover $H = \mathrm{Aut}(\Omega/M)$, then $\mathrm{Aut}(\Omega/M_1)$ is a normal subgroup of H, and $\sigma \mapsto \sigma|M_1$ maps $H/\mathrm{Aut}(\Omega/M_1)$ isomorphically onto a dense subgroup of $\mathrm{Aut}(M_1/M)$.

PROOF. See 6.3 of Shimura, *Introduction to the arithmetic theory of automorphic functions*. Princeton, 1971. □

Exercises

9-1 Find the centralizer of complex conjugation in $\text{Aut}(\mathbb{C}/\mathbb{Q})$.

Review Exercises

A-1 Let p be a prime number, and let m and n be positive integers.

(a) Give necessary and sufficient conditions on m and n for $\mathbb{F}_{p^n}$ to have a subfield isomorphic with $\mathbb{F}_{p^m}$. Prove your answer.

(b) If there is such a subfield, how many subfields isomorphic with $\mathbb{F}_{p^m}$ are there, and why?

A-2 Show that the Galois group of the splitting field F of $X^3 - 7$ over $\mathbb{Q}$ is isomorphic to S_3, and exhibit the fields between $\mathbb{Q}$ and F. Which of the fields between $\mathbb{Q}$ and F are normal over $\mathbb{Q}$?

A-3 Prove that the two fields $\mathbb{Q}[\sqrt{7}]$ and $\mathbb{Q}[\sqrt{11}]$ are not isomorphic.

A-4 (a) Prove that the multiplicative group of all nonzero elements in a finite field is cyclic.

(b) Construct explicitly a field of order 9, and exhibit a generator for its multiplicative group.

A-5 Let X be transcendental over a field F, and let E be a subfield of $F(X)$ properly containing F. Prove that X is algebraic over E.

A-6 Prove as directly as you can that if ζ is a primitive pth root of 1, p prime, then the Galois group of $\mathbb{Q}[\zeta]$ over $\mathbb{Q}$ is cyclic of order $p - 1$.

A-7 Let G be the Galois group of the polynomial $X^5 - 2$ over $\mathbb{Q}$.

(a) Determine the order of G.

(b) Determine whether G is abelian.

(c) Determine whether G is solvable.

A-8   (a) Show that every field homomorphism from $\mathbb{R}$ to $\mathbb{R}$ is bijective.

(b) Prove that $\mathbb{C}$ is isomorphic to infinitely many different subfields of itself.

A-9 Let F be a field with 16 elements. How many roots in F does each of the following polynomials have? $X^3 - 1$; $X^4 - 1$; $X^{15} - 1$; $X^{17} - 1$.

A-10 Find the degree of a splitting field of the polynomial $(X^3 - 5)(X^3 - 7)$ over $\mathbb{Q}$.

A-11 Find the Galois group of the polynomial $X^6 - 5$ over each of the fields $\mathbb{Q}$ and $\mathbb{R}$.

A-12 The coefficients of a polynomial $f(X)$ are algebraic over a field F. Show that $f(X)$ divides some nonzero polynomial $g(X)$ with coefficients in F.

A-13 Let $f(X)$ be a polynomial in $F[X]$ of degree n, and let E be a splitting field of f. Show that $[E:F]$ divides $n!$.

A-14 Find a primitive element for the field $\mathbb{Q}[\sqrt{3}, \sqrt{7}]$ over $\mathbb{Q}$, i.e., an element such that $\mathbb{Q}[\sqrt{3}, \sqrt{7}] = \mathbb{Q}[\alpha]$.

A-15 Let G be the Galois group of $(X^4 - 2)(X^3 - 5)$ over $\mathbb{Q}$.

(a) Give a set of generators for G, as well as a set of defining relations.

(b) What is the structure of G as an abstract group (is it cyclic, dihedral, alternating, symmetric, etc.)?

A-16 Let F be a finite field of characteristic $\neq 2$. Prove that $X^2 = -1$ has a solution in F if and only if $|F| \equiv 1 \mod 4$.

A-17 Let E be the splitting field over $\mathbb{Q}$ of $(X^2 - 2)(X^2 - 5)(X^2 - 7)$. Find an element α in E such that $E = \mathbb{Q}[\alpha]$. (You must prove that $E = \mathbb{Q}[\alpha]$.)

A-18 Let E be a Galois extension of F with Galois group S_n, $n > 1$ not prime. Let H_1 be the subgroup of S_n of elements fixing 1, and let H_2 be the subgroup generated by the cycle $(123\ldots n)$. Let $E_i = E^{H_i}$, $i = 1, 2$. Find the degrees of E_1, E_2, $E_1 \cap E_2$, and $E_1 E_2$ over F. Show that there exists a field M such that $F \subset M \subset E_2$, $M \neq F$, $M \neq E_2$, but that no such field exists for E_1.

A-19 Let ζ be a primitive 12th root of 1 over $\mathbb{Q}$. How many fields are there strictly between $\mathbb{Q}[\zeta^3]$ and $\mathbb{Q}[\zeta]$.

A-20 For the polynomial $X^3 - 3$, find explicitly its splitting field over $\mathbb{Q}$ and elements that generate its Galois group.

A-21 Let $E = \mathbb{Q}[\zeta]$, $\zeta^5 = 1$, $\zeta \neq 1$. Show that $i \notin E$, and that if $L = E[i]$, then -1 is a norm from L to E. Here $i = \sqrt{-1}$.

A-22 Let E be an extension of F, and let Ω be an algebraic closure of E. Let $\sigma_1, \ldots, \sigma_n$ be distinct F-isomorphisms $E \to \Omega$.

(a) Show that $\sigma_1, \ldots, \sigma_n$ are linearly dependent over Ω.

(b) Show that $[E:F] \geq m$.

(c) Let F have characteristic $p > 0$, and let L be a subfield of Ω containing E and such that $a^p \in E$ for all $a \in L$. Show that each σ_i has a unique extension to a homomorphism $\sigma_i' : L \to \Omega$.

A-23 Identify the Galois group of the splitting field F of $X^4 - 3$ over $\mathbb{Q}$. Determine the number of quadratic subfields.

A-24 Let F be a subfield of a finite field E. Prove that the trace map $T = \mathrm{Tr}_{E/F}$ and the norm map $N = \mathrm{Nm}_{E/F}$ of E over F both map E *onto* F. (You may quote basic properties of finite fields and the trace and norm.)

A-25 Prove or disprove by counterexample.

(a) If L/F is an extension of fields of degree 2, then there is an automorphism σ of L such that F is the fixed field of σ.

(b) The same as (a) except that L is also given to be finite.

A-26 A finite Galois extension L of a field K has degree 8100. Show that there is a field F with $K \subset F \subset L$ such that $[F:K] = 100$.

A-27 An algebraic extension L of a field K of characteristic 0 is generated by an element θ that is a root of both of the polynomials $X^3 - 1$ and $X^4 + X^2 + 1$. Given that $L \neq K$, find the minimal polynomial of θ.

A-28 Let $F/\mathbb{Q}$ be a Galois extension of degree 3^n, $n \geq 1$. Prove that there is a chain of fields
$$\mathbb{Q} = F_0 \subset F_1 \subset \cdots F_n = F$$
such that for every i, $0 \leq i \leq n-1$, $[F_{i+1} : F_i] = 3$.

A-29 Let L be the splitting field over $\mathbb{Q}$ of an equation of degree 5 with distinct roots. Suppose that L has an automorphism that fixes three of these roots while interchanging the other two and also an automorphism $\alpha \neq 1$ of order 5.

(a) Prove that the group of automorphisms of L is the symmetric group on 5 elements.

(b) How many proper subfields of L are normal extensions of $\mathbb{Q}$? For each such field F, what is $[F:\mathbb{Q}]$?

A-30 If L/K is a separable algebraic field extension of finite degree d, show that the number of fields between K and L is at most $2^{d!}$. [This is far from best possible. See math.stackexchange.com, question 522976.]

A-31 Let K be the splitting field over $\mathbb{Q}$ of $X^5 - 1$. Describe the Galois group $\mathrm{Gal}(K/\mathbb{Q})$ of K over $\mathbb{Q}$, and show that K has exactly one subfield of degree 2 over $\mathbb{Q}$, namely, $\mathbb{Q}[\zeta + \zeta^4]$, $\zeta \neq 1$ a root of $X^5 - 1$. Find the minimal polynomial of $\zeta + \zeta^4$ over $\mathbb{Q}$. Find $\mathrm{Gal}(L/\mathbb{Q})$ when L is the splitting field over $\mathbb{Q}$ of

(a) $(X^2 - 5)(X^5 - 1)$;

(b) $(X^2 + 3)(X^5 - 1)$.

A-32 Let Ω_1 and Ω_2 be algebraically closed fields of transcendence degree 5 over $\mathbb{Q}$, and let $\alpha\colon \Omega_1 \to \Omega_2$ be a homomorphism (in particular, $\alpha(1) = 1$). Show that α is a bijection. (State carefully all theorems you use.)

A-33 Find the group of $\mathbb{Q}$-automorphisms of the field $k = \mathbb{Q}[\sqrt{-3}, \sqrt{-2}]$.

A-34 Prove that the polynomial $f(X) = X^3 - 5$ is irreducible over the field $\mathbb{Q}[\sqrt{7}]$. If L is the splitting field of $f(X)$ over $\mathbb{Q}[\sqrt{7}]$, prove that the Galois group of $L/\mathbb{Q}[\sqrt{7}]$ is isomorphic to S_3. Prove that there must exist a subfield K of L such that the Galois group of L/K is cyclic of order 3.

A-35 Identify the Galois group G of the polynomial $f(X) = X^5 - 6X^4 + 3$ over F, when (a) $F = \mathbb{Q}$ and when (b) $F = \mathbb{F}_2$. In each case, if E is the splitting field of $f(X)$ over F, determine how many fields K there are such that $E \supset K \supset F$ with $[K:F] = 2$.

A-36 Let K be a field of characteristic p, say with p^n elements, and let θ be the automorphism of K that maps every element to its pth power. Show that there exists an automorphism α of K such that $\theta\alpha^2 = 1$ if and only if n is odd.

A-37 Describe the splitting field and Galois group, over $\mathbb{Q}$, of the polynomial $X^5 - 9$.

A-38 Suppose that E is a Galois field extension of a field F such that $[E:F] = 5^3 \cdot (43)^2$. Prove that there exist fields K_1 and K_2 lying strictly between F and E with the following properties: (i) each K_i is a Galois extension of F; (ii) $K_1 \cap K_2 = F$; and (iii) $K_1 K_2 = E$.

A-39 Let $F = \mathbb{F}_p$ for some prime p. Let m be a positive integer not divisible by p, and let K be the splitting field of $X^m - 1$. Find $[K:F]$ and prove that your answer is correct.

A-40 Let F be a field of 81 elements. For each of the following polynomials $g(X)$, determine the number of roots of $g(X)$ that lie in F: $X^{80} - 1$, $X^{81} - 1$, $X^{88} - 1$.

A-41 Describe the Galois group of the polynomial $X^6 - 7$ over $\mathbb{Q}$.

A-42 Let K be a field of characteristic $p > 0$ and let $F = K(u, v)$ be a field extension of degree p^2 such that $u^p \in K$ and $v^p \in K$. Prove that K is not finite, that F is not a simple extension of K, and that there exist infinitely many intermediate fields $F \supset L \supset K$.

A-43 Find the splitting field and Galois group of the polynomial $X^3 - 5$ over the field $\mathbb{Q}[\sqrt{2}]$.

A-44 For every prime p, find the Galois group over $\mathbb{Q}$ of the polynomial $X^5 - 5p^4 X + p$.

A-45 Factorize $X^4 + 1$ over each of the finite fields (a) $\mathbb{F}_5$; (b) $\mathbb{F}_{25}$; and (c) $\mathbb{F}_{125}$. Find its splitting field in each case.

A-46 Let $\mathbb{Q}[\alpha]$ be a field of finite degree over $\mathbb{Q}$. Assume that there is a $q \in \mathbb{Q}$, $q \neq 0$, such that $|\rho(\alpha)| = q$ for all homomorphisms $\rho : \mathbb{Q}[\alpha] \to \mathbb{C}$. Show that the set of roots of the minimal polynomial of α is the same as that of q^2/α. Deduce that there exists an automorphism σ of $\mathbb{Q}[\alpha]$ such that

(a) $\sigma^2 = 1$ and

(b) $\rho(\sigma\gamma) = \overline{\rho(\gamma)}$ for all $\gamma \in \mathbb{Q}[\alpha]$ and $\rho : \mathbb{Q}[\alpha] \to \mathbb{C}$.

A-47 Let F be a field of characteristic zero, and let p be a prime number. Suppose that F has the property that all irreducible polynomials $f(X) \in F[X]$ have degree a power of p ($1 = p^0$ is allowed). Show that every equation $g(X) = 0$, $g \in F[X]$, is solvable by extracting radicals.

A-48 Let $K = \mathbb{Q}[\sqrt{5}, \sqrt{-7}]$ and let L be the splitting field over $\mathbb{Q}$ of $f(X) = X^3 - 10$.

(a) Determine the Galois groups of K and L over $\mathbb{Q}$.

(b) Decide whether K contains a root of f.

(c) Determine the degree of the field $K \cap L$ over $\mathbb{Q}$.

[Assume all fields are subfields of $\mathbb{C}$.]

A-49 Find the splitting field (over $\mathbb{F}_p$) of $X^{p^r} - X \in \mathbb{F}_p[X]$, and deduce that $X^{p^r} - X$ has an irreducible factor $f \in \mathbb{F}_p[X]$ of degree r. Let $g(X) \in \mathbb{Z}[X]$ be a monic polynomial that becomes equal to $f(X)$ when its coefficients are read modulo p. Show that $g(X)$ is irreducible in $\mathbb{Q}[X]$.

A-50 Let E be the splitting field of $X^3 - 51$ over $\mathbb{Q}$. List all the subfields of E, and find an element γ of E such that $E = \mathbb{Q}[\gamma]$.

A-51 Let $k = \mathbb{F}_{1024}$ be the field with 1024 elements, and let K be an extension of k of degree 2. Prove that there is a unique automorphism σ of K of order 2 which leaves k elementwise fixed and determine the number of elements of $K^\times$ such that $\sigma(x) = x^{-1}$.

A-52 Let F and E be finite fields of the same characteristic. Prove or disprove these statements:

(a) There is a ring homomorphism of F into E if and only if $|E|$ is a power of $|F|$.

(b) There is an injective group homomorphism of the multiplicative group of F into the multiplicative group of E if and only if $|E|$ is a power of $|F|$.

A-53 Let L/K be an algebraic extension of fields. Prove that L is algebraically closed if every polynomial over K factors completely over L.

A-54 Let K be a field, and let $M = K(X)$, X an indeterminate. Let L be an intermediate field different from K. Prove that M is finite-dimensional over L.

A-55 Let $\theta_1, \theta_2, \theta_3$ be the roots of the polynomial $f(X) = X^3 + X^2 - 9X + 1$.

(a) Show that the θ_i are real, nonrational, and distinct.

(b) Explain why the Galois group of $f(X)$ over $\mathbb{Q}$ must be either A_3 or S_3. Without carrying it out, give a brief description of a method for deciding which it is.

(c) Show that the rows of the matrix

$$\begin{pmatrix} 3 & 9 & 9 & 9 \\ 3 & \theta_1 & \theta_2 & \theta_3 \\ 3 & \theta_2 & \theta_3 & \theta_1 \\ 3 & \theta_3 & \theta_1 & \theta_2 \end{pmatrix}$$

are pairwise orthogonal; compute their lengths, and compute the determinant of the matrix.

A-56 Let E/K be a Galois extension of degree $p^2 q$ where p and q are primes, $q < p$ and q not dividing $p^2 - 1$. Prove that:

(a) there exist intermediate fields L and M such that $[L:K] = p^2$ and $[M:K] = q$;

(b) such fields L and M must be Galois over K; and

(c) the Galois group of E/K must be abelian.

A-57 Let ζ be a primitive 7th root of 1 (in $\mathbb{C}$).

(a) Prove that $1 + X + X^2 + X^3 + X^4 + X^5 + X^6$ is the minimal polynomial of ζ over $\mathbb{Q}$.

(b) Find the minimal polynomial of $\zeta + \frac{1}{\zeta}$ over $\mathbb{Q}$.

A-58 Find the degree over $\mathbb{Q}$ of the Galois closure K of $\mathbb{Q}[2^{\frac{1}{4}}]$ and determine the isomorphism class of $\mathrm{Gal}(K/\mathbb{Q})$.

A-59 Let p, q be distinct positive prime numbers, and consider the extension $K = \mathbb{Q}[\sqrt{p}, \sqrt{q}] \supset \mathbb{Q}$.

(a) Prove that the Galois group is isomorphic to $C_2 \times C_2$.

(b) Prove that every subfield of K of degree 2 over $\mathbb{Q}$ is of the form $\mathbb{Q}[\sqrt{m}]$ where $m \in \{p, q, pq\}$.

(c) Show that there is an element $\gamma \in K$ such that $K = \mathbb{Q}[\gamma]$.

Appendix **B**

Two-hour Examination

1. (a) Let σ be an automorphism of a field E. If $\sigma^4 = 1$ and

$$\sigma(\alpha) + \sigma^3(\alpha) = \alpha + \sigma^2(\alpha) \qquad \text{all } \alpha \in E,$$

show that $\sigma^2 = 1$.

(b) Let p be a prime number and let a, b be rational numbers such that $a^2 + pb^2 = 1$. Show that there exist rational numbers c, d such that $a = \frac{c^2 - pd^2}{c^2 + pd^2}$ and $b = \frac{2cd}{c^2 + pd^2}$.

2. Let $f(X)$ be an irreducible polynomial of degree 4 in $\mathbb{Q}[X]$, and let $g(X)$ be the resolvent cubic of f. What is the relation between the Galois group of f and that of g? Find the Galois group of f if

(a) $g(X) = X^3 - 3X + 1$;

(b) $g(X) = X^3 + 3X + 1$.

3. (a) How many monic irreducible factors does $X^{255} - 1 \in \mathbb{F}_2[X]$ have, and what are their degrees.

(b) How many monic irreducible factors does $X^{255} - 1 \in \mathbb{Q}[X]$ have, and what are their degrees?

4. Let E be the splitting field of $(X^5 - 3)(X^5 - 7) \in \mathbb{Q}[X]$. What is the degree of E over $\mathbb{Q}$? How many proper subfields of E are there that are not contained in the splitting fields of both $X^5 - 3$ and $X^5 - 7$?

[You may assume that 7 is not a 5th power in the splitting field of $X^5 - 3$.]

5. Consider an extension $\Omega \supset F$ of fields. Define $a \in \Omega$ to be *F-constructible* if it is contained in a field of the form

$$F[\sqrt{a_1}, \ldots, \sqrt{a_n}], \qquad a_i \in F[\sqrt{a_1}, \ldots, \sqrt{a_{i-1}}].$$

Assume Ω is a finite Galois extension of F and construct a field E, $F \subset E \subset \Omega$, such that every $a \in \Omega$ is E-constructible and E is minimal with this property.

6. Let Ω be an extension field of a field F. Show that every F-homomorphism $\Omega \to \Omega$ is an isomorphism provided:

  (a) Ω is algebraically closed, and

  (b) Ω has finite transcendence degree over F.

    Can either of the conditions (i) or (ii) be dropped? (Either prove, or give a counterexample.)

You should prove all answers. You may use results proved in class or in the notes, but you should indicate clearly what you are using.

Possibly useful facts: The discriminant of $X^3 + aX + b$ is $-4a^3 - 27b^2$ and $2^8 - 1 = 255 = 3 \times 5 \times 17$.

Appendix **C**

Solutions to the Exercises

These solutions fall somewhere between hints and complete solutions. Students were expected to write out complete solutions.

1-1. Similar to Example 1.28.

1-2. Verify that 3 is not a square in $\mathbb{Q}[\sqrt{2}]$, and so $[\mathbb{Q}[\sqrt{2}, \sqrt{3}]:\mathbb{Q}] = 4$.

1-3. (a) Apply the division algorithm, to get $f(X) = q(X)(X - a) + r(X)$ with $r(X)$ constant, and put $X = a$ to find $r = f(a)$.
(c) Use that factorization in $F[X]$ is unique, or use induction on the degree of f.
(d) If G had two cyclic factors C and C' whose orders were divisible by a prime p, then G would have (at least) p^2 elements of order dividing p. This does not happen, and it follows that G is cyclic.
(e) The elements of order m in $F^\times$ are the roots of the polynomial $X^m - 1$, and so there are at most m of them. Hence every finite subgroup G of $F^\times$ satisfies the condition in (d).

1-4. Note that it suffices to construct $\alpha = \cos\frac{2\pi}{7}$, and that $[\mathbb{Q}[\alpha]:\mathbb{Q}] = \frac{7-1}{2} = 3$, and so its minimal polynomial has degree 3 (see Example 3.22). There is a standard method (once taught in high schools) for solving cubics using the equation

$$\cos 3\theta = 4\cos^3\theta - 3\cos\theta.$$

By "completing the cube", reduce the cubic to the form $X^3 - pX - q$. Then construct a square root a of $\frac{4p}{3}$, so that $a^2 = \frac{4p}{3}$. Let 3θ be the angle such that $\cos 3\theta = \frac{4q}{a^3}$, and use the angle trisector to construct $\cos\theta$. From the displayed equation, we find that $\alpha = a\cos\theta$ is a root of $X^3 - pX - q$. For a geometric construction, see sx93476.

171

1-5. Let f_1 be an irreducible factor of f in $E[X]$, and let (L, α) be a stem field for f_1 over E. Then $m \mid [L:F]$ because $L \supset E$ (1.20). But $f(\alpha) = 0$, and so $(F[\alpha], \alpha)$ is a stem field for f over F, which implies that $[F[\alpha]:F] = n$. Now $n \mid [L:F]$ because $L \supset F[\alpha]$. We deduce that $[L:F] = mn$ and $[L:E] = n$. But $[L:E] = \deg(f_1)$, and so $f_1 = f$.

1-6. The polynomials $f(X) - 1$ and $f(X) + 1$ have only finitely many roots, and so there exists an $n \in \mathbb{Z}$ such that $f(n) \neq \pm 1$. Let p be a prime dividing $f(n)$. Then $f(n) = 0$ modulo p, and so f has a root in $\mathbb{F}_p$. Thus it is not irreducible in $\mathbb{F}_p[X]$.

1-7. It is easy to see that R is ring, and so it remains to show that every nonzero element $a + b\alpha + c\alpha^2$ has an inverse in R. Let $f(X) = X^3 - 2$ and $g(X) = cX^2 + bX + a$. As f is irreducible and $\deg(g) < \deg(f)$, f and g are relatively prime. Therefore Euclid's algorithm gives polynomials $u(X)$ and $v(X)$ with $\deg v < 3$ such that $u(X)f(X) + v(X)g(X) = 1$. On putting $X = \alpha$ in this equation, we find that $v(\alpha)g(\alpha) = 1$, i.e., $v(\alpha)$ is inverse to $g(\alpha) = a + b\alpha + c\alpha^2$. Alternatively, R is an integral domain (being a subring of $\mathbb{C}$), and so (1.23) shows that R is a field.

1-8. This is Problem 4, p. 3, in Borevich and Shafarevich. Number theory. Academic Press, 1966.

2-1. (a) is obvious, as is the "only if" in (b). For the "if" note that for any $a \in S(E)$, $a \notin F^2$, $E \approx F[X]/(X^2 - a)$.

(c) Take $E_i = \mathbb{Q}[\sqrt{p_i}]$ with p_i the ith prime. Check that p_i is the only prime that becomes a square in E_i. For this use that $(a + b\sqrt{p})^2 \in \mathbb{Q} \implies 2ab = 0$.

(d) Every field of characteristic p contains (an isomorphic copy of) $\mathbb{F}_p$, and so we are looking at the quadratic extensions of $\mathbb{F}_p$. The homomorphism $a \mapsto a^2 : \mathbb{F}_p^\times \to \mathbb{F}_p^\times$ has kernel $\{\pm 1\}$, and so its image has index 2 in $\mathbb{F}_p^\times$. Thus the only possibility for $S(E)$ is $\mathbb{F}_p^\times$, and so there is at most one E (up to $\mathbb{F}_p$-isomorphism). To get one, take $E = F[X]/(X^2 - a)$, $a \notin \mathbb{F}_p^2$.

2-2. (a) If α is a root of $f(X) = X^p - X - a$ (in some splitting field), then the remaining roots are $\alpha + 1, \ldots, \alpha + p - 1$, which obviously lie in whichever field contains α. Moreover, they are distinct. Suppose that, in $F[X]$,

$$f(X) = (X^r + a_1 X^{r-1} + \cdots + a_r)(X^{p-r} + \cdots), \quad 0 < r < p.$$

Then $-a_1$ is a sum of r of the roots of f, $-a_1 = r\alpha + d$ some $d \in \mathbb{Z} \cdot 1_F$, and it follows that $\alpha \in F$.

(b) As 0 and 1 are not roots of $X^p - X - 1$ in $\mathbb{F}_p$ it cannot have p distinct roots in $\mathbb{F}_p$, and so (a) implies that $X^p - X - 1$ is irreducible in $\mathbb{F}_p[X]$ and hence also in $\mathbb{Z}[X]$ and $\mathbb{Q}[X]$ (see 1.18, 1.13).

2-3. Let α be the real 5th root of 2. Eisenstein's criterion shows that $X^5 - 2$ is irreducible in $\mathbb{Q}[X]$, and so $\mathbb{Q}[\sqrt[5]{2}]$ has degree 5 over $\mathbb{Q}$. The remaining roots of $X^5 - 2$ are $\zeta\alpha, \zeta^2\alpha, \zeta^3\alpha, \zeta^4\alpha$, where ζ is a primitive 5th root of 1. It follows that the subfield of $\mathbb{C}$ generated by the roots of $X^5 - 2$ is $\mathbb{Q}[\zeta, \alpha]$. The degree of $\mathbb{Q}[\zeta, \alpha]$ is 20, since it must be divisible by $[\mathbb{Q}[\zeta]:\mathbb{Q}] = 4$ and $[\mathbb{Q}[\alpha]:\mathbb{Q}] = 5$.

2-4. The splitting field of the first polynomial is $\mathbb{F}_p$ because $X^{p^m} - 1 = (X - 1)^{p^m}$. For the second, see Proposition 4.20.

2-5. If $f(X) = \prod(X - \alpha_i)^{m_i}$, $\alpha_i \neq \alpha_j$, then

$$f'(X) = \sum m_i \frac{f(X)}{X - \alpha_i}$$

and so $d(X) = \prod_{m_i > 1}(X - \alpha_i)^{m_i - 1}$. Therefore $g(X) = \prod(X - \alpha_i)$.

2-6. From (2.20) we know that either f is separable or $f(X) = f_1(X^p)$ for some polynomial f_1. Clearly f_1 is also irreducible. If f_1 is not separable, it can be written $f_1(X) = f_2(X^p)$. Continue in the way until you arrive at a separable polynomial. For the final statement, note that $g(X) = \prod(X - a_i)$, $a_i \neq a_j$, and so $f(X) = g(X^{p^e}) = \prod(X - \alpha_i)^{p^e}$ with $\alpha_i^{p^e} = a_i$.

3-1. Let σ and τ be automorphisms of $F(X)$ given by $\sigma(X) = -X$ and $\tau(X) = 1 - X$. Then σ and τ fix X^2 and $X^2 - X$ respectively, and so $\sigma\tau$ fixes $E \overset{\text{def}}{=} F(X) \cap F(X^2 - X)$. But $\sigma\tau X = 1 + X$, and so $(\sigma\tau)^m(X) = m + X$. Thus $\text{Aut}(F(X)/E)$ is infinite, which implies that $[F(X):E]$ is infinite (otherwise $F(X) = E[\alpha_1, \ldots, \alpha_n]$; an E-automorphism of $F(X)$ is determined by its values on the α_i, and its value on α_i is a root of the minimal polynomial of α_i). If E contains a polynomial $f(X)$ of degree $m > 0$, then $[F(X):E] \leq [F(X):F(f(X))] = m$ — contradiction.

3-2. Since $1 + \zeta + \cdots + \zeta^{p-1} = 0$, we have $\alpha + \beta = -1$. If $i \in H$, then $iH = H$ and $i(G \smallsetminus H) = G \smallsetminus H$, and so α and β are fixed by H. If $j \in G \smallsetminus H$, then $jH = G \smallsetminus H$ and $j(G \smallsetminus H) = H$, and so $j\alpha = \beta$ and $j\beta = \alpha$. Hence $\alpha\beta \in \mathbb{Q}$, and α and β are the roots of $X^2 + X + \alpha\beta$.

Note that
$$\alpha\beta = \sum_{i,j} \zeta^{i+j}, \quad i \in H, \quad j \in G \smallsetminus H.$$

How many times do we have $i + j = 0$? If $i + j = 0$, then $-1 = i^{-1}j$, which is a nonsquare; conversely, if -1 is a nonsquare, take $i = 1$ and $j = -1$ to get $i + j = 0$. Hence

$$i + j = 0 \text{ some } i \in H, \ j \in G \smallsetminus H \iff -1 \text{ is a square mod } p$$
$$\iff p \equiv -1 \mod 4.$$

If we do have a solution to $i + j = 0$, we get all solutions by multiplying it through by the $\frac{p-1}{2}$ squares. So in the sum for $\alpha\beta$ we see 1 a total of $\frac{p-1}{2}$ times when $p \equiv 3 \mod 4$ and not at all if $p \equiv 1 \mod 4$. In either case, the remaining terms add to a rational number, which implies that each power of ζ occurs the same number of times.

Thus for $p \equiv 1 \mod 4$, $\alpha\beta = -(\frac{p-1}{2})^2/(p-1) = -\frac{p-1}{4}$; the polynomial satisfied by α and β is $X^2 + X - \frac{p-1}{4}$, whose roots are $(-1 \pm \sqrt{1+p-1})/2$; the fixed field of H is $\mathbb{Q}[\sqrt{p}]$.

For $p \equiv -1 \mod 4$,

$$\alpha\beta = \frac{p-1}{2} + (-1)\left(\left(\frac{p-1}{2}\right)^2 - \frac{p-1}{2}\right)/(p-1) = \frac{p-1}{2} - \frac{p-3}{4} = \frac{p+1}{4};$$

the polynomial is $X^2 + X + \frac{p+1}{4}$, with roots $(-1 \pm \sqrt{1-p-1})/2$; the fixed field of H is $\mathbb{Q}[\sqrt{-p}]$. See also sx984457.

3-3. (a) It is easy to see that M is Galois over $\mathbb{Q}$ with Galois group $\langle \sigma, \tau \rangle$:

$$\begin{cases} \sigma\sqrt{2} = -\sqrt{2} \\ \sigma\sqrt{3} = \sqrt{3} \end{cases} \qquad \begin{cases} \tau\sqrt{2} = \sqrt{2} \\ \tau\sqrt{3} = -\sqrt{3} \end{cases}.$$

(b) We have

$$\frac{\sigma\alpha^2}{\alpha^2} = \frac{2-\sqrt{2}}{2+\sqrt{2}} = \frac{(2-\sqrt{2})^2}{4-2} = \left(\frac{2-\sqrt{2}}{\sqrt{2}}\right)^2 = (\sqrt{2}-1)^2,$$

i.e., $\sigma\alpha^2 = ((\sqrt{2}-1)\alpha)^2$. Thus, if $\alpha \in M$, then $\sigma\alpha = \pm(\sqrt{2}-1)\alpha$, and

$$\sigma^2\alpha = (-\sqrt{2}-1)(\sqrt{2}-1)\alpha = -\alpha;$$

as $\sigma^2\alpha = \alpha \neq 0$, this is impossible. Hence $\alpha \notin M$, and so $[E:\mathbb{Q}] = 8$. Extend σ to an automorphism (also denoted σ) of E. Again $\sigma\alpha = \pm(\sqrt{2}-1)\alpha$ and $\sigma^2\alpha = -\alpha$, and so $\sigma^2 \neq 1$. Now $\sigma^4\alpha = \alpha$, $\sigma^4|M = 1$, and so we can conclude that σ has order 4. After possibly replacing σ with its inverse, we may suppose that $\sigma\alpha = (\sqrt{2}-1)\alpha$.

Repeat the above argument with τ: $\frac{\tau\alpha^2}{\alpha^2} = \frac{3-\sqrt{3}}{3+\sqrt{3}} = \left(\frac{3-\sqrt{3}}{\sqrt{6}}\right)^2$, and so we can extend τ to an automorphism of L (also denoted τ) with $\tau\alpha = \frac{3-\sqrt{3}}{\sqrt{6}}\alpha$. The order of τ is 4.

Finally compute that

$$\sigma\tau\alpha = \frac{3-\sqrt{3}}{-\sqrt{6}}(\sqrt{2}-1)\alpha; \quad \tau\sigma\alpha = (\sqrt{2}-1)\frac{3-\sqrt{3}}{\sqrt{6}}\alpha.$$

Hence $\sigma\tau \neq \tau\sigma$, and $\mathrm{Gal}(E/\mathbb{Q})$ has two noncommuting elements of order 4. Since it has order 8, it must be the quaternion group.

See also sx983458.

3-5. Let $G = \mathrm{Aut}(E/F)$. Then E is Galois over E^G with Galois group G, and so $|G| = [E:E^G]$. Now $[E:F] = [E:E^G][E^G:F] = |G|[E^G:F]$.

4-1. The splitting field is the smallest field containing all mth roots of 1. Hence it is $\mathbb{F}_{p^n}$ where n is the smallest positive integer such that $m_0 | p^n - 1$, $m = m_0 p^r$, where p is prime and does not divide m_0.

4-2. We have $X^4 - 2X^3 - 8X - 3 = (X^3 + X^2 + 3X + 1)(X - 3)$, and $g(X) = X^3 + X^2 + 3X + 1$ is irreducible over $\mathbb{Q}$ (use 1.11), and so its Galois group is either A_3 or S_3. Either check that its discriminant is not a square or, more simply, show by examining its graph that $g(X)$ has only one real root, and hence its Galois group contains a transposition (cf. the proof of 4.16).

4-3. Eisenstein's criterion shows that $X^8 - 2$ is irreducible over $\mathbb{Q}$, and so $[\mathbb{Q}[\alpha]:\mathbb{Q}] = 8$ where α is a positive 8th root of 2. As usual for polynomials of this type, the splitting field is $\mathbb{Q}[\alpha, \zeta]$ where ζ is any primitive 8th root of 1. For example, ζ can be taken to be $\frac{1+i}{\sqrt{2}}$, which lies in $\mathbb{Q}[\alpha, i]$. It follows that the splitting field is $\mathbb{Q}[\alpha, i]$. Clearly $\mathbb{Q}[\alpha, i] \neq \mathbb{Q}[\alpha]$, because $\mathbb{Q}[\alpha]$, unlike i, is contained in $\mathbb{R}$, and so $[\mathbb{Q}[\alpha, i]:\mathbb{Q}[\alpha]] = 2$. Therefore the degree is $2 \times 8 = 16$.

4-4. Find an extension L/F with Galois group S_4, and let E be the fixed field of $S_3 \subset S_4$. There is no subgroup strictly between S_n and S_{n-1}, because such a subgroup would be transitive and contain an $(n-1)$-cycle and a transposition, and so would equal S_n. We can take $E = L^{S_3}$. More specifically, we can take L to be the splitting field of $X^4 - X + 2$ over $\mathbb{Q}$ and E to be the subfield generated by a root of the polynomial (see 3.27).

4-5. Type: "Factor$(X^{343} - X)$ mod 7;" and discard the 7 factors of degree 1.

4-6. Type "galois($X^6 + 2X^5 + 3X^4 + 4X^3 + 5X^2 + 6X + 7$);". It is the group $\mathrm{PGL}_2(\mathbb{F}_5)$ (group of invertible 2×2 matrices over $\mathbb{F}_5$ modulo scalar matrices) which has order 120. Alternatively, note that there are the following factorizations: mod 3, irreducible; mod 5 (deg 3)(deg 3); mod 13 (deg 1)(deg 5); mod 19, (deg 1)^2(deg 4); mod 61 (deg 1)^2(deg 2)^2; mod 79, (deg 2)^3. Thus the Galois group has elements of type:

$$6, \quad 3+3, \quad 1+5, \quad 1+1+4, \quad 1+1+2+2, \quad 2+2+2.$$

No element of type 2, 3, $3+2$, or $4+2$ turns up by factoring modulo any of the first 400 primes (or, so I have been told). This suggests it is the group $T14$ in the tables in Butler and McKay, which is indeed $\mathrm{PGL}_2(\mathbb{F}_5)$.

4-7. $\Longleftarrow$: Condition (a) implies that G_f contains a 5-cycle, condition (b) implies that $G_f \subset A_5$, and condition (c) excludes A_5. That leaves D_5 and C_5 as the only possibilities (see, for example, Jacobson, *Basic Algebra I*, p. 305, Ex 6). The derivative of f is $5X^4 + a$, which has at most 2 real zeros, and so (from its graph) we see that f can have at most 3 real zeros. Thus complex conjugation acts as an element of order 2 on the splitting field of f, and this shows that we must have $G_f = D_5$.

$\Longrightarrow$: Regard D_5 as a subgroup of S_5 by letting it act on the vertices of a regular pentagon—all subgroups of S_5 isomorphic to D_5 look like this one. If $G_f = D_5$, then (a) holds because D_5 is transitive, (b) holds because $D_5 \subset A_5$, and (c) holds because D_5 is solvable.

4-8. Suppose that f is irreducible of degree n. Then f has no root in a field $\mathbb{F}_{p^m}$ with $m < n$, which implies (a). However, every root α of f lies in $\mathbb{F}_{p^n}$, and so $\alpha^{p^n} - \alpha = 0$. Hence $(X - \alpha) | (X^{p^n} - X)$, which implies (b) because f has no multiple roots.

Conversely, suppose that (a) and (b) hold. It follows from (b) that all roots of f lie in $\mathbb{F}_{p^n}$. Suppose that f had an irreducible factor g of degree $m < n$. Then every root of g generates $\mathbb{F}_{p^m}$, and so $\mathbb{F}_{p^m} \subset \mathbb{F}_{p^n}$. Consequently, m divides n, and so m divides n/p_i for some i. But then g divides both f and $X^{p^{n/p_i}} - X$, contradicting (a). Thus f is irreducible.

4-9. Let a_1, a_2 be conjugate nonreal roots, and let a_3 be a real root. Complex conjugation defines an element σ of the Galois group of f switching a_1 and a_2 and fixing a_3. On the other hand, because f is irreducible, its Galois group acts transitively on its roots, and so there is a τ such that $\tau(a_3) = a_1$.

Now

$$a_3 \overset{\tau}{\mapsto} a_1 \overset{\sigma}{\mapsto} a_2$$
$$a_3 \overset{\sigma}{\mapsto} a_3 \overset{\tau}{\mapsto} a_1.$$

This statement is false for reducible polynomials — consider for example $f(X) = (X^2 + 1)(X - 1)$.

4-10. See mo113794.

5-1. For $a = 1$, this is the polynomial $\Phi_5(X)$, whose Galois group is cyclic of order 4.

For $a = 0$, $f(X) = X(X^3 + X^2 + X + 1) = X(X + 1)(X^2 + 1)$, whose Galois group is cyclic of order 2.

For $a = 12$, $f(X) = (X^2 - 2X + 3)(X^2 + 3X + 4)$, whose Galois group is V_4 (the one not sitting inside A_4).

For $a = -4$, $f(X) = (X - 1)(X^3 + 2X^2 + 3X + 4)$. The cubic does not have $\pm 1, \pm 2$, or ± 4 as roots, and so it is irreducible in $\mathbb{Q}[X]$. Hence its Galois group is S_3 or A_3. Modulo 13, $f(X) = (X - 1)(X - 2)(X^2 + 4X - 2)$, and so the Galois group contains a 2-cycle by Dedekind's theorem. Therefore, it is S_3. Alternatively, use that the discriminant of the cubic is -200, which is not a square. Note that, because 2 divides the discriminant, we cannot use Dedekind's theorem with $p = 2$.

For a general a, the resolvent cubic is

$$g(X) = X^3 - X^2 + (1 - 4a)X + 3a - 1.$$

For $a = -1$, $f = X^4 + X^3 + X^2 + X - 1$ is irreducible modulo 2, and so it is irreducible over $\mathbb{Q}$. The resolvant cubic is $g = X^3 - X^2 + 5X - 4$, which is irreducible. Moreover

$$g'(x) = 3x^2 - 2x + 5 = 3(x - \tfrac{1}{3})^2 + 4\tfrac{2}{3} > 0, \text{ all } x,$$

and so g has exactly one real root. Hence the Galois group of g is S_3, and it follows that the Galois group of f is S_4.

Thus we have found the following Galois groups (in S_4): C_2, C_4, V_4 ($\not\subseteq A_4$), S_3, S_4. This seems to be all. The discriminant of f is $256a^3 - 203a^2 + 88a - 16$. If a is odd, this is odd, and we can apply Dedekind's theorem with $p = 2$ to show that the Galois group contains a 2-cycle or a 4-cycle, and so $1, A_3, A_4, V_4$ are not possible. In the general case, the discriminant is not a square, and so the Galois group is not contained in A_4.

Showing that the discriminant is not a square is equivalent to solving for integral points on the elliptic curve $Y^2 = 256X^3 - 203X^2 + 88X - 16$. The substitution $X \mapsto X/2^{16}$, $Y \mapsto Y/2^{20}$ turns this into the equation

$$Y^2 = X^3 - 51968X^2 + 1476395008X - 17592186044416.$$

According to PARI this has no nonzero rational points, and so the discriminant cannot be a square. (I thank Ivan Ip for his help with this solution.)

5-2. We have $\mathrm{Nm}(a + ib) = a^2 + b^2$. Hence $a^2 + b^2 = 1$ if and only if $a + ib = \frac{s+it}{s-it}$ for some $s, t \in \mathbb{Q}$ (Hilbert's Theorem 90). The rest is easy.

5-3. The degree $[\mathbb{Q}[\zeta_n] : \mathbb{Q}] = \varphi(n)$, ζ_n a primitive nth root of 1, and $\varphi(n) \to \infty$ as $n \to \infty$.

5-4. Let $\alpha_1, \ldots, \alpha_n$ be the roots of f in E, and let H_i be the subgroup of $\mathrm{Gal}(E/F)$ fixing $F[\alpha_i]$. As $E \neq F[\alpha_i]$, $H_i \neq 1$. As f is irreducible, $\mathrm{Gal}(E/F)$ acts transitively on $\{\alpha_1, \ldots, \alpha_n\}$, and hence on $\{F[\alpha_1], \ldots, F[\alpha_n]\}$, which is a set with more than one element. The H_i are all conjugate, and so none is normal.

9-1. If some element centralizes complex conjugation, then it must preserve the real numbers as a set. Now, since any automorphism of the real numbers preserves the set of squares, it must preserve the order; and hence be continuous. Since $\mathbb{Q}$ is fixed, this implies that the real numbers are fixed pointwise. It follows that any element that centralizes complex conjugation must be the identity or the complex conjugation itself. See mo121083, Andreas Thom.

A-1. (a) Need that $m|n$, because

$$n = [\mathbb{F}_{p^n} : \mathbb{F}_p] = [\mathbb{F}_{p^n} : \mathbb{F}_{p^m}] \cdot [\mathbb{F}_{p^m} : \mathbb{F}_p] = [\mathbb{F}_{p^n} : \mathbb{F}_{p^m}] \cdot m.$$

Use Galois theory to show there exists one, for example. (b) Only one; it consists of all the solutions of $X^{p^m} - X = 0$.

A-2. The polynomial is irreducible by Eisenstein's criterion. The polynomial has only one real root, and therefore complex conjugation is a transposition in G_f. This proves that $G_f \approx S_3$. The discriminant is $-1323 = -3^3 7^2$. Only the subfield $\mathbb{Q}[\sqrt{-3}]$ is normal over $\mathbb{Q}$. The subfields $\mathbb{Q}[\sqrt[3]{7}]$, $\mathbb{Q}[\zeta \sqrt[3]{7}]$ $\mathbb{Q}[\zeta^2 \sqrt[3]{7}]$ are not normal over $\mathbb{Q}$. [The discriminant of $X^3 - a$ is $-27a^2 = -3(3a)^2$.]

A-3. The prime 7 becomes a square in the first field, but 11 does not: $(a + b\sqrt{7})^2 = a^2 + 7b^2 + 2ab\sqrt{7}$, which lies in $\mathbb{Q}$ only if $ab = 0$. Hence the rational numbers that become squares in $\mathbb{Q}[\sqrt{7}]$ are those that are already squares or lie in $7\mathbb{Q}^{\times 2}$.

A-4. (a) See Exercise 3.

(b) Let $F = \mathbb{F}_3[X]/(X^2 + 1)$. Modulo 3

$$X^8 - 1 = (X - 1)(X + 1)(X^2 + 1)(X^2 + X + 2)(X^2 + 2X + 2).$$

Take α to be a root of $X^2 + X + 2$.

A-5. Since $E \neq F$, E contains an element $\frac{f}{g}$ with the degree of f or $g > 0$. Now

$$f(T) - \frac{f(X)}{g(X)} g(T)$$

is a nonzero polynomial having X as a root.

A-6. Use Eisenstein to show that $X^{p-1} + \cdots + 1$ is irreducible, etc. Done in class.

A-7. The splitting field is $\mathbb{Q}[\zeta, \alpha]$ where $\zeta^5 = 1$ and $\alpha^5 = 2$. It is generated by $\sigma = (12345)$ and $\tau = (2354)$, where $\sigma\alpha = \zeta\alpha$ and $\tau\zeta = \zeta^2$. The group has order 20. It is not abelian (because $\mathbb{Q}[\alpha]$ is not Galois over $\mathbb{Q}$), but it is solvable (its order is < 60).

A-8. (a) A homomorphism $\alpha : \mathbb{R} \to \mathbb{R}$ acts as the identity map on $\mathbb{Z}$, hence on $\mathbb{Q}$, and it maps positive real numbers to positive real numbers, and therefore preserves the order. Hence, for each real number a,

$$\{r \in \mathbb{Q} \mid a < r\} = \{r \in \mathbb{Q} \mid \alpha(a) < r\},$$

which implies that $\alpha(a) = a$.

(b) Choose a transcendence basis A for $\mathbb{C}$ over $\mathbb{Q}$. Because it is infinite, there is a bijection $\alpha : A \to A'$ from A onto a proper subset. Extend α to an isomorphism $\mathbb{Q}(A) \to \mathbb{Q}(A')$, and then extend it to an isomorphism $\mathbb{C} \to \mathbb{C}'$ where $\mathbb{C}'$ is the algebraic closure of $\mathbb{Q}(A')$ in $\mathbb{C}$.

A-9. The group $F^{\times}$ is cyclic of order 15. It has 3 elements of order dividing 3, 1 element of order dividing 4, 15 elements of order dividing 15, and 1 element of order dividing 17.

A-10. If E_1 and E_2 are Galois extensions of F, then $E_1 E_2$ and $E_1 \cap E_2$ are Galois over F, and $\mathrm{Gal}(E_1 E_2 / F)$ is the fibred product of $\mathrm{Gal}(E_1/F)$ and

$\mathrm{Gal}(E_2/F)$ over $\mathrm{Gal}(E_1 \cap E_2/F)$:

$$
\begin{array}{ccc}
\mathrm{Gal}(E_1 E_2/F) & \longrightarrow & \mathrm{Gal}(E_1/F) \\
\downarrow & & \downarrow \\
\mathrm{Gal}(E_2/F) & \longrightarrow & \mathrm{Gal}(E_1 \cap E_2/F).
\end{array}
$$

In this case, $E_1 \cap E_2 = \mathbb{Q}[\zeta]$ where ζ is a primitive cube root of 1. The degree is 18.

A-11. Over $\mathbb{Q}$, the splitting field is $\mathbb{Q}[\alpha, \zeta]$ where $\alpha^6 = 5$ and $\zeta^3 = 1$ (because $-\zeta$ is then a primitive 6th root of 1). The degree is 12, and the Galois group is D_6 (generators $(26)(35)$ and (123456)).

Over $\mathbb{R}$, the Galois group is C_2.

A-12. Let the coefficients of f be $a_1,\ldots,a_n$ — they lie in the algebraic closure Ω of F. Let $g(X)$ be the product of the minimal polynomials over F of the roots of f in Ω.

Alternatively, the coefficients will lie in some finite extension E of F, and we can take the norm of $f(X)$ from $E[X]$ to $F[X]$.

A-13. If f is separable, $[E:F] = (G_f:1)$, which is a subgroup of S_n. Etc..

A-14. $\sqrt{3} + \sqrt{7}$ will do.

A-15. The splitting field of $X^4 - 2$ is $E_1 = \mathbb{Q}[i,\alpha]$ where $\alpha^4 = 2$; it has degree 8, and Galois group D_4. The splitting field of $X^3 - 5$ is $E_2 = \mathbb{Q}[\zeta,\beta]$; it has degree 6, and Galois group D_3. The Galois group is the product (they could only intersect in $\mathbb{Q}[\sqrt{3}]$, but $\sqrt{3}$ does not become a square in E_1).

A-16. The multiplicative group of F is cyclic of order $q-1$. Hence it contains an element of order 4 if and only if $4|q-1$.

A-17. Take $\alpha = \sqrt{2} + \sqrt{5} + \sqrt{7}$.

A-18. We have $E_1 = E^{H_1}$, which has degree n over F, and $E_2 = E^{<1\cdots n>}$, which has degree $(n-1)!$ over F, etc.. This is really a problem in group theory posing as a problem in field theory.

A-19. We have $\mathbb{Q}[\zeta] = \mathbb{Q}[i,\zeta']$ where ζ' is a primitive cube root of 1 and $\pm i = \zeta^3$ etc..

A-20. The splitting field is $\mathbb{Q}[\zeta, \sqrt[3]{3}]$, and the Galois group is S_3.

A-21. Use that

$$(\zeta + \zeta^4)(1 + \zeta^2) = \zeta + \zeta^4 + \zeta^3 + \zeta$$

A-22. (a) is Dedekind's theorem. (b) is Artin's theorem 3.4. (c) is O.K. because $X^p - a^p$ has a unique root in Ω.

A-23. The splitting field is $\mathbb{Q}[i, \alpha]$ where $\alpha^4 = 3$, and the Galois group is D_4 with generators (1234) and (13) etc..

A-24. From Hilbert's theorem 90, we know that the kernel of the map $N: E^\times \to F^\times$ consists of elements of the form $\frac{\sigma\alpha}{\alpha}$. The map $E^\times \to E^\times$, $\alpha \mapsto \frac{\sigma\alpha}{\alpha}$, has kernel $F^\times$. Therefore the kernel of N has order $\frac{q^m - 1}{q - 1}$, and hence its image has order $q - 1$. There is a similar proof for the trace — I do not know how the examiners expected you to prove it.

A-25. (a) is false—could be inseparable. (b) is true—could not be inseparable.

A-26. Apply the Sylow theorem to see that the Galois group has a subgroup of order 81. Now the Fundamental Theorem of Galois theory shows that F exists.

A-27. The greatest common divisor of the two polynomials over $\mathbb{Q}$ is $X^2 + X + 1$, which must therefore be the minimal polynomial for θ.

A-28. Theorem on p-groups plus the Fundamental Theorem of Galois Theory.

A-29. It was proved in class that S_p is generated by an element of order p and a transposition (4.15). There is only one F, and it is quadratic over $\mathbb{Q}$.

A-30. Let $L = K[\alpha]$. The splitting field of the minimal polynomial of α has degree at most $d!$, and a set with $d!$ elements has at most $2^{d!}$ subsets. [Of course, this bound is much too high: the subgroups are very special subsets. For example, they all contain 1 and they are invariant under $a \mapsto a^{-1}$.]

A-31. The Galois group is $(\mathbb{Z}/5\mathbb{Z})^\times$, which cyclic of order 4, generated by 2.

$$(\zeta + \zeta^4) + (\zeta^2 + \zeta^3) = -1, \quad (\zeta + \zeta^4)(\zeta^2 + \zeta^3) = -1.$$

(a) Omit.

(b) Certainly, the Galois group is a product $C_2 \times C_4$.

A-32. Let $a_1, \ldots, a_5$ be a transcendence basis for $\Omega_1/\mathbb{Q}$. Their images are algebraically independent, therefore they are a maximal algebraically independent subset of Ω_2, and therefore they form a transcendence basis, etc..

A-33. $C_2 \times C_2$.

A-34. If $f(X)$ were reducible over $\mathbb{Q}[\sqrt{7}]$, it would have a root in it, but it is irreducible over $\mathbb{Q}$ by Eisenstein's criterion. The discriminant is -675, which is not a square in $\mathbb{R}$, much less $\mathbb{Q}[\sqrt{7}]$.

A-35. (a) Should be $X^5 - 6X^4 + 3$. The Galois group is S_5, with generators (12) and (12345) — it is irreducible (Eisenstein) and (presumably) has exactly 2 nonreal roots. (b) It factors as $(X + 1)(X^4 + X^3 + X^2 + X + 1)$. Hence the splitting field has degree 4 over $\mathbb{F}_2$, and the Galois group is cyclic.

A-36. This is really a theorem in group theory, since the Galois group is a cyclic group of order n generated by θ. If n is odd, say $n = 2m + 1$, then $\alpha = \theta^m$ does.

A-37. It has order 20, generators (12345) and (2354).

A-38. Take K_1 and K_2 to be the fields corresponding to the Sylow 5 and Sylow 43 subgroups. Note that of the possible numbers $1, 6, 11, 16, 21, \dots$ of Sylow 5-subgroups, only 1 divides 43. There are 1, 44, 87, ... subgroups of
....

A-39. See Exercise 14.

A-40. The group $F^\times$ is cyclic of order 80; hence 80, 1, 8.

A-41. It's D_6, with generators (26)(35) and (123456). The polynomial is irreducible by Eisenstein's criterion, and its splitting field is $\mathbb{Q}[\alpha, \zeta]$ where $\zeta \neq 1$ is a cube root of 1.

A-42. Example 5.5.

A-43. Omit.

A-44. It's irreducible by Eisenstein. Its derivative is $5X^4 - 5p^4$, which has the roots $X = \pm p$. These are the max and mins, $X = p$ gives negative; $X = -p$ gives positive. Hence the graph crosses the x-axis 3 times and so there are 2 imaginary roots. Hence the Galois group is S_5.

A-45. Its roots are primitive 8th roots of 1. It splits completely in $\mathbb{F}_{25}$. (a) $(X^2 + 2)(X^2 + 3)$.

A-46. $\rho(\alpha)\overline{\rho(\alpha)} = q^2$, and $\rho(\alpha)\rho(\frac{q^2}{\alpha}) = q^2$. Hence $\rho(\frac{q^2}{\alpha})$ is the complex conjugate of $\rho(\alpha)$. Hence the automorphism induced by complex conjugation is independent of the embedding of $\mathbb{Q}[\alpha]$ into $\mathbb{C}$.

A-47. The argument that proves the Fundamental Theorem of Algebra, shows that its Galois group is a p-group. Let E be the splitting field of $g(X)$, and let H be the Sylow p-subgroup. Then $E^H = F$, and so the Galois group is a p-group.

A-48. (a) $C_2 \times C_2$ and S_3. (b) No. (c). 1

A-49. Omit.

A-50. Omit.

A-51. $1024 = 2^{10}$. Want $\sigma x \cdot x = 1$, i.e., $Nx = 1$. They are the elements of the form $\frac{\sigma x}{x}$; have

$$1 \longrightarrow k^\times \longrightarrow K^\times \xrightarrow{\ x \mapsto \frac{\sigma x}{x}\ } K^\times.$$

Hence the number is $2^{11}/2^{10} = 2$.

A-52. Pretty standard. False; true.

A-53. Omit.

A-54. Similar to a previous problem.

A-55. Omit.

A-56. This is really a group theory problem disguised as a field theory problem.

A-57. (a) Prove it's irreducible by apply Eisenstein to $f(X + 1)$. (b) See example worked out in class.

A-58. It's D_4, with generators (1234) and (12).

A-59. Omit.

SOLUTIONS FOR THE EXAM.

1. (a) Let σ be an automorphism of a field E. If $\sigma^4 = 1$ and

$$\sigma(\alpha) + \sigma^3(\alpha) = \alpha + \sigma^2(\alpha) \qquad \text{all } \alpha \in E,$$

show that $\sigma^2 = 1$.

If $\sigma^2 \neq 1$, then $1, \sigma, \sigma^2, \sigma^3$ are distinct automorphisms of E, and hence are linearly independent (Dedekind 5.14) — contradiction. [If $\sigma^2 = 1$, then the condition becomes $2\sigma = 2$, so either $\sigma = 1$ or the characteristic is 2 (or both).]

(b) Let p be a prime number and let a, b be rational numbers such that $a^2 + pb^2 = 1$. Show that there exist rational numbers c, d such that $a = \frac{c^2 - pd^2}{c^2 + pd^2}$ and $b = \frac{2cd}{c^2 + pd^2}$.

Apply Hilbert's Theorem 90 to $\mathbb{Q}[\sqrt{-p}]$.

2. Let $f(X)$ be an irreducible polynomial of degree 4 in $\mathbb{Q}[X]$, and let $g(X)$ be the resolvent cubic of f. What is the relation between the Galois group of f and that of g? Find the Galois group of f if

(a) $g(X) = X^3 - 3X + 1$;

(b) $g(X) = X^3 + 3X + 1$.

We have $G_g = G_f / G_f \cap V$, where $V = \{1, (12)(34), \ldots\}$. The two cubic polynomials are irreducible, because their only possible roots are ± 1. From their discriminants, one finds that the first has Galois group A_3 and the second S_3. Because $f(X)$ is irreducible, $4 | (G_f : 1)$ and it follows that $G_f = A_4$ and S_4 in the two cases.

3. (a) How many monic irreducible factors does $X^{255} - 1 \in \mathbb{F}_2[X]$ have, and what are their degrees?

Its roots are the nonzero elements of $\mathbb{F}_{2^8}$, which has the following subfields $\mathbb{F}_{2^4} \supset \mathbb{F}_{2^2} \supset \mathbb{F}_2$. There are $256 - 16$ elements not in $\mathbb{F}_{16}$, and their minimal polynomials all have degree 8. Hence there are 30 factors of degree 8, 3 of degree 4, and 1 each of degrees 2 and 1.

(b) How many monic irreducible factors does $X^{255} - 1 \in \mathbb{Q}[X]$ have, and what are their degrees?

Obviously, $X^{255} - 1 = \prod_{d | 255} \Phi_d = \Phi_1 \Phi_3 \Phi_5 \Phi_{15} \cdots \Phi_{255}$, and we showed in class that the Φ_d are irreducible. They have degrees $1, 2, 4, 8, 16, 32, 64, 128$.

4. Let E be the splitting field of $(X^5 - 3)(X^5 - 7) \in \mathbb{Q}[X]$. What is the degree of E over $\mathbb{Q}$? How many proper subfields of E are there that are not contained in the splitting fields of both $X^5 - 3$ and $X^5 - 7$?

The splitting field of $X^5 - 3$ is $\mathbb{Q}[\zeta, \alpha]$, which has degree 5 over $\mathbb{Q}[\zeta]$ and 20 over $\mathbb{Q}$. The Galois group of $X^5 - 7$ over $\mathbb{Q}[\zeta, \alpha]$ is (by ...) a subgroup of a cyclic group of order 5, and hence has order 1 or 5. Since 7 is not a 5th power in $\mathbb{Q}[\zeta, \alpha]$, it must be 5. Thus $[E : \mathbb{Q}] = 100$, and

$$G = \mathrm{Gal}(E/\mathbb{Q}) = (C_5 \times C_5) \rtimes C_4.$$

We want the nontrivial subgroups of G not containing $C_5 \times C_5$. The subgroups of order 5 of $C_5 \times C_5$ are lines in $(\mathbb{F}_5)^2$, and hence $C_5 \times C_5$ has $6 + 1 = 7$ proper subgroups. All are normal in G. Each subgroup of $C_5 \times C_5$ is of the form $H \cap (C_5 \times C_5)$ for exactly 3 subgroups H of G corresponding to the three possible images in $G/(C_5 \times C_5) = C_4$. Hence we have 21 subgroups of G not containing $C_5 \times C_5$, and 20 nontrivial ones. Typical fields: $\mathbb{Q}[\alpha]$, $\mathbb{Q}[\alpha, \cos \frac{2\pi}{5}]$, $\mathbb{Q}[\alpha, \zeta]$.

[You may assume that 7 is not a 5th power in the splitting field of $X^5 - 3$.]

5. Consider an extension $\Omega \supset F$ of fields. Define $\alpha \in \Omega$ to be F-*constructible* if it is contained in a field of the form

$$F[\sqrt{a_1}, \ldots, \sqrt{a_n}], \qquad a_i \in F[\sqrt{a_1}, \ldots, \sqrt{a_{i-1}}].$$

Assume Ω is a finite Galois extension of F and construct a field E, $F \subset E \subset \Omega$, such that every $a \in \Omega$ is E-constructible and E is minimal with this property.

Suppose E has the required property. From the primitive element theorem, we know $\Omega = E[a]$ for some a. Now a E-constructible $\implies [\Omega : E]$ is a power of 2. Take $E = \Omega^H$, where H is the Sylow 2-subgroup of $\mathrm{Gal}(\Omega/F)$.

6. Let Ω be an extension field of a field F. Show that every F-homomorphism $\Omega \to \Omega$ is an isomorphism provided:

  (a) Ω is algebraically closed, and

  (b) Ω has finite transcendence degree over F.

  Can either of the conditions (i) or (ii) be dropped? (Either prove, or give a counterexample.)

  Let A be a transcendence basis for Ω/F. Because $\sigma : \Omega \to \Omega$ is injective, $\sigma(A)$ is algebraically independent over F, and hence (because it has the right number of elements) is a transcendence basis for Ω/F. Now $F[\sigma A] \subset \sigma\Omega \subset \Omega$. Because Ω is algebraic over $F[\sigma A]$ and $\sigma\Omega$ is algebraically closed, the two are equal. Neither condition can be dropped. E.g., $\mathbb{C}(X) \to \mathbb{C}(X)$, $X \mapsto X^2$. E.g., $\Omega =$ the algebraic closure of $\mathbb{C}(X_1, X_2, X_3, \ldots)$, and consider an extension of the map $X_1 \mapsto X_2$, $X_2 \mapsto X_3, \ldots$.

Index